London Mathematical Society Lecture Note Series. 1

PETER HILTON

General Cohomology Theory and K-Theory

Course given at the University of São Paulo in the summer of 1968 under the auspices of the Instituto de Pesquisas Matemáticas, Universidade de São Paulo.

CAMBRIDGE AT THE UNIVERSITY PRESS 1971

CAMBRIDGE UNIVERSITY PRESS
Cambridge, New York, Melbourne, Madrid, Cape Town, Singapore, São Paulo

Cambridge University Press
The Edinburgh Building, Cambridge CB2 8RU, UK

Published in the United States of America by Cambridge University Press, New York

www.cambridge.org
Information on this title: www.cambridge.org/9780521079761

First published 1971
Re-issued in this digitally printed version 2008

A catalogue record for this publication is available from the British Library

Library of Congress Catalogue Card Number: 74–127239

ISBN 978-0-521-07976-1 paperback

Contents

Introduction

In the summer of 1968 I received an invitation to give a series of lectures on some topic in algebraic topology at the University of São Paulo, Brazil. The level was to be approximately that of a second year graduate course (at Cornell University, for example); that is to say, the audience would consist of people versed in ordinary homology and cohomology theory and in homotopy theory. It further appeared that the course would consist of nine lectures, each of one and a half hours' duration, so that the topic had to be one in which a satisfactory 'pay-off' could be achieved in a relatively short time.

The topic of general cohomology theory, with special reference to K-theory, suggested itself very naturally. The existence of a comprehensive and readable literature, both original papers and books, made it justifiable to highlight some of the more sensational achievements of K-theory without giving details of all proofs. * Moreover, by presenting K-theory as just one - though, to be sure, one of the most exciting - of the 'new' cohomology theories, it was natural to give some attention to properties of general cohomology theories and their relation to ordinary cohomology. The study of the entire category of cohomology theories and their interrelations is itself an area of active research and I was

---

* The following two books constituted the basic reference works for the course, as they do for these notes: M. F. Atiyah, K-theory, Benjamin; D. Husemoller, Fibre Bundles, McGraw-Hill.

particularly concerned to place my audience in the position of being able to tackle open questions in this area; on the other hand, K-theory itself is now a well-established tool in topology and is probably not itself a promising topic for research among new initiates.

The leitmotif of the course was the presentation of two connections between ordinary and general cohomology, namely, the generalized Atiyah-Hirzebruch spectral sequence and the character. Specializing to (complex) K-theory, one obtains the Atiyah-Hirzebruch spectral sequence and the Chern character and these two tools were used to prove Adams' celebrated theorem on the non-existence of elements of Hopf invariant 1 in $\pi_{2n-1}(S^n)$, $n \neq 2, 4, 8$. Since Adams' original proof involved the development of the subtle Adams spectral sequence relating cohomology to stable homotopy and a deep study of secondary cohomology operations (Adams' final paper in Annals of Mathematics occupied 80 pages!), it is a signal triumph of K-theory to produce a proof which only requires a knowledge of certain very natural primary operations in K-theory. Applications of Adams' theory were given to obtain results in classical linear algebra.

A more detailed description of these course notes is as follows. In Chapter I we define general cohomology theories, both reduced and unreduced, and list certain elementary results. A feature of the presentation here is that the theory is presented as an absolute theory and the corresponding relative theory is deduced from it. This makes for a great gain in simplicity and reduces the number of axioms to three, the suspension axiom replacing the usual coboundary homomorphism; on the other hand, the more usual (relative) axiom system may be more suitable for certain categories of topological spaces (e.g. compacta). The important notion of a representable theory is described in this chapter.

In Chapter II a general theory of spectral sequences is developed. This chapter could be largely ignored (except insofar as it establishes notation) by anybody familiar with spectral sequence theory. The treatment is based on that of Eckmann-Hilton [Exact couples in an abelian category, Journ. of Algebra (1966), pp. 38-87], but is considerably more elementary in that only abelian groups are considered and very explicit constructions are made of the various groups and homomorphisms which arise in the functorial passage from exact couples to spectral sequences. It is not at all claimed that this more elementary approach is conceptually simple; however, it appeared at the time that the audience was unfamiliar with more abstract categorical reasoning, and these notes, intended as a faithful record of the course, reflect the audience's preference. The reader who prefers the more categorical approach should consult the paper by Eckmann-Hilton. Not everything in this chapter is directly applicable to the material of Chapter III, but the structure of the chapter was adapted to the purpose of this very particular application.

Chapter III describes the generalized Atiyah-Hirzebruch spectral sequence. For a given finite-dimensional complex X this spectral sequence passes from ordinary cohomology $H^p(X;\check{h}^{q-p})$ with coefficients in the $(q-p)^{th}$ component of the coefficients of the theory h, converging to the graded group associated with $h^q(X)$, suitably filtered. The filtration is given by $F^p h^q(X) = \ker(h^q(X) \to h^q(X_{p-1}))$, where X_{p-1} is the (p-1)-skeleton of X. Moreover, the convergence is finite and the filtration is finite; precisely if $\dim X = k$, then $E_k = E_{k+1} = \ldots = E_\infty$ in the spectral sequence and $F^{k+1}h(X) = 0$, $F^0 h(X) = h(X)$. Various conclusions are drawn from the spectral sequence; in particular it is shown that if Q is the group of rationals then

$$h^n(X) \otimes Q = \bigoplus_{p+q=n} H^p(X;Q) \otimes \check{h}^q .$$

The natural transformation $h \to h \otimes Q$ is called the character of the theory h. As the formula above shows, $h \otimes Q$ depends only on ordinary cohomology (with rational coefficients) and the torsion-free part of the coefficient group of h.

In Chapter IV K-theory is described using vector bundles over the given space X. Since all the emphasis here is on complex K-theory, the vector bundles are themselves complex, but, of course, real vector bundles are also presented. It is shown how the Grothendieck group K(X) of equivalence classes of vector bundles over X breaks up naturally into a direct sum $\tilde{K}(X) \oplus Z$, the second component being the dimension of the 'fibre' over the base point. Then $\tilde{K}(X)$ may be identified with the set of based homotopy classes of maps, $X \to B_U \times Z$, where B_U is the classifying space for the 'big' unitary group U. Then Bott periodicity, asserting a homotopy equivalence between ΩU and $B_U \times Z$, leads to the definition of the reduced cohomology theory $\tilde{K}^n(X)$ and the free cohomology theory $K^n(X)$. Moreover, the tensor product of bundles leads to a commutative ring structure in K(X) with $\tilde{K}(X)$ as an ideal. The Grothendieck group is particularly simple in this case due to the fact that, to any bundle ξ, there exists a bundle η such that the Whitney sum $\xi \oplus \eta$ is a trivial bundle. It follows easily from this that $\tilde{K}(X)$ is naturally isomorphic to the set of stable equivalence classes of vector bundles over X.

In Chapter V the Chern character is defined using the classical theory of Chern classes for a complex vector bundle and it is identified with the character defined in Chapter III from the spectral sequence. The Adams operations in K-theory are introduced and their relation to the Chern character is elucidated with the aid of the so-called splitting principle; this principle asserts that, given any vector bundle ξ over X there is a space X_ξ and a map $f: X_\xi \to X$ such that f^* is a monomorphism in ordinary integral cohomology (and therefore also in cohomology over Q) and $f^*\xi$ is a sum of line-bundles. Properties of the Chern character in the special case where the homology of X is torsion-free, together with the Adams operations, are then used to prove the celebrated Adams theorem on the non-existence of elements of Hopf invariant 1. The actual theorem proved asserts that if X is a finite complex such that $H^*(X;Z) = Z[a]/a^3$ with $\dim a = n$, then $n = 2, 4$ or 8. Consequences of this theorem then close the chapter.

The course itself culminated in a discussion of the problem of extending cohomology theories from certain categories of topological spaces to larger categories. The material in this final section of the course was a digest of an article which has recently been published, namely: Peter Hilton, On the construction of cohomology theories, Rend. di Matem. (Roma), 1968, 219-232. This article, which was itself based on a talk given at the Istituto Matematico, Universita di Roma, dealing with the work of A. Deleanu and the author, is here reproduced as an appendix to these notes, by kind permission of Professor B. Segre.

It should be emphasized that this article only initiates the discussion of the extension problem and leaves open many questions. It is thus to be hoped that further work along these lines will prove fruitful. Of course, other topics would have served equally well (or better) to round off the course and provide the audience with the stimulus to undertake original work on general cohomology.

Many acknowledgements are due. I am indebted to my friend Frank Adams for the inspiration provided by his seminal work on general cohomology theories, K-theory, the Hopf invariant problem, the vector fields problem, and a host of other major contributions to algebraic topology. I am indebted to my friend Beno Eckmann for allowing me to borrow freely from the notes of the course, Cohomologie et Classes Caractéristiques, which he gave at a symposium held under the auspices of C. I. M. E. in 1966. I am indebted to my friend Aristide Deleanu for allowing me to publish the appendix. I am indebted to my friends and colleagues Carlos de Lyra and Renzo Piccinini for preparing these notes with great care, skillfully establishing order out of my somewhat informal presentations; they are in no way to be held responsible for the residual imperfections of this text. Above all, however, I am grateful to the Instituto de Pesquisas Matematicas for inviting me to São Paulo to give this course, to Professor de Lyra for the fine way in which he made arrangements to secure the best possible conditions for my visit, and to the many friends and colleagues, acquired during those very pleasant weeks, whose enthusiasm for mathematics was only matched by their kindness and hospitality.

Peter Hilton

Cornell University, Ithaca, New York
January 1969

Note added in proof.

Since the preparation of these notes, further progress has been made in connection with the theory described in the Appendix. This progress is reported in the following articles:

A. Deleanu and P. J. Hilton, 'On the generalized Čech construction of cohomology theories', Battelle Research Report No. 28, 1969.

A. Deleanu and P. J. Hilton, 'On extensions of cohomology theories and Serre classes of groups', Battelle Research Report No. 34, 1970.

1. General Cohomology Theories

We shall denote by $\mathscr{C}$ the category of finite-dimensional cell-complexes with base point; the morphisms are base-point preserving maps from one based space to another. Furthermore, homotopies will be assumed to preserve base-points.

For the definition of a general cohomology theory, we need the (reduced) suspension functor in $\mathscr{C}$, a covariant functor Σ from $\mathscr{C}$ to $\mathscr{C}$ defined as follows. Denote by I the unit interval $[0, 1]$ and let $\dot{I}$ be the subspace $\{0, 1\}$ of I. Let $(X, x_0) \in \mathscr{C}$; then ΣX is defined to be the quotient space $X \times I/(X \times \dot{I} \cup x_0 \times I)$ with the obvious base-point. If $f:(X, x_0) \to (Y, y_0)$ is a map of $\mathscr{C}$, define $\Sigma f: \Sigma X \to \Sigma Y$ by observing that the map $(x, t) \to (f(x), t)$ is compatible with passage to the quotient and thus induces a map Σf.

We also need the construction known as the mapping cone. Let $CX = X \times I/(X \times 1 \cup x_0 \times I)$ be the reduced cone on X with X embedded in CX by $x \to (x, 0)$; and let $f:X \to Y$ be a map of $\mathscr{C}$. Define the mapping cone C_f of f as the quotient space of the topological sum $CX + Y$ by the following identifications: x is identified to $f(x) \in Y$ for all $x \in X$. In particular, if $u:X \to \{x_0\}$ is the constant map, C_u is just ΣX. There is an evident embedding $i:Y \subseteq C_f$. Also if $f:X \to Y$ is an inclusion of a subcomplex, there is a natural homotopy equivalence $C_f \to Y/X$.

Let X, Y $\in \mathscr{C}$ have base-points x_0 and y_0 respectively. The spaces $X \vee Y$ and $X \wedge Y$ are defined as follows: $X \vee Y = X \times \{y_0\} \cup \{x_0\} \times Y$ with base-point (x_0, y_0) and topology given by the natural imbedding $X \vee Y \subset X \times Y$; the space $X \wedge Y$ is defined to be the quotient space $X \times Y/X \vee Y$. The operations $\vee$ and $\wedge$ are associative and commutative in the category $\mathscr{C}$;

furthermore $\wedge$ is distributive with respect to $\vee$ (Spanier, E.: Function Spaces and Duality, Annals of Math. 70 (1959); 338-378).

In particular, if $X = S^1$, the standard unit 1-sphere with base-point, we have the natural homeomorphism $S^1 \wedge Y \approx \Sigma Y$. It is also not difficult to check that $S^n \wedge S^m \approx S^{n+m}$. By iteration one gets $S^n \wedge Y \approx \Sigma^n(Y)$.

A (reduced) cohomology theory is a family $h = \{h^n; n \in Z\}$ of contravariant functors from the category $\mathscr{C}$ to the category of abelian groups and homomorphisms, together with a family of natural transformations $\sigma = \{\sigma^n; n \in Z\}$, $\sigma^n : h^n \to h^{n+1}\Sigma$ subject to the following axioms:

(1. 1) (Homotopy axiom) If $f \simeq g$, then $h^n(f) = h^n(g)$ for every $n \in$ (or, as we may write, $f^* = g^*$);

(1. 2) (Suspension axiom) σ^n is a natural equivalence for every $n \in Z$;

(1. 3) (Exactness axiom) Given $X \xrightarrow{f} Y \xrightarrow{i} C_f$, the sequence of abelian groups and homomorphisms

$$h^n(X) \xleftarrow{f^*} h^n(Y) \xleftarrow{i^*} h^n(C_f)$$

is exact for every n.

Associated to every map $f: X \to Y$, a long exact sequence arises as follows. Apply axiom (1. 3) successively to the sequence

$$X \xrightarrow{f} Y \xrightarrow{i} C_f \xrightarrow{j} C_i \simeq \Sigma X \xrightarrow{\Sigma f} \Sigma Y \longrightarrow \dots,$$

which arises using the homotopy equivalence $C_i \simeq C_f/Y = \Sigma X$. We get an exact sequence (using (1. 1) and (1. 2))

$$\begin{array}{ccccccccc} h^n(X) & \xleftarrow{f^*} & h^n(Y) & \xleftarrow{i^*} & h^n(C_f) & \xleftarrow{j^*} & h^n(\Sigma X) & \xleftarrow{(\Sigma f)^*} & h^n(\Sigma Y) \\ & & & & & & \sigma^{n-1}\uparrow \cong & & \uparrow \cong \\ & & & & & & h^{n-1}(X) & \xleftarrow{f^*} & h^{n-1}(Y) \dots \end{array}$$

or

(1. 4) $$\dots \longleftarrow h^n(X) \xleftarrow{f^*} h^n(Y) \xleftarrow{i^*} h^n(C_f) \xleftarrow{\delta} h^{n-1}(X) \xleftarrow{f^*} h^{n-1}(Y) \longleftarrow \dots,$$

where $\delta = j^* \sigma^{n-1}$.

If the map $f:X \to Y$ is an imbedding, we define $h^n(Y, X) = h^n(C_f) = h^n(Y/X)$. Using the isomorphism $h^n(\Sigma X) \cong h^{n-1}(X)$ of axiom (1.2), we get the expected long exact sequence of the pair (Y, X):

$$(1.5) \quad \ldots \longleftarrow h^n(X) \longleftarrow h^n(Y) \longleftarrow h^n(Y, X) \longleftarrow h^{n-1}(X) \longleftarrow \ldots$$

Thus our axiom system implies the usual Eilenberg-Steenrod axioms except for the dimension axiom (see definition below).

As a first example of a general cohomology theory we have the usual cohomology theory H; in fact, this theory satisfies one more axiom besides axioms (1.1), (1.2) and (1.3), namely the dimension axiom, which asserts that for a 0-sphere S^0, $h^n(S^0) = 0$ if $n \neq 0$. Actually, if a cohomology theory satisfies also the dimension axiom, then we have uniqueness in the category $\mathscr{C}_f$ of finite cell-complexes. That is, there is only ordinary cohomology theory (with specified coefficients). We will give a proof of this later.

We list a few more examples of cohomology theories.

(1) Given any theory h and an integer k, a new theory ${}^k h$ can be trivially obtained by setting $({}^k h)^n = h^{n+k}$. This is called 'suspending the theory h'.

(2) Given any space $Z \in \mathscr{C}$ and a theory h, we define a new theory h_Z by setting $h_Z^n(X) = h^n(X \wedge Z)$, for every $X \in \mathscr{C}$, $n \in Z$.

To verify axiom (1.2), recall that $\Sigma X = S^1 \wedge X$ and use the associativity $S^1 \wedge (X \wedge Z) \approx (S^1 \wedge X) \wedge Z = (\Sigma X) \wedge Z$. Axiom (1.3) is satisfied because, given $f:X \to Y$, the spaces $C_f \wedge Z$ and $C_{f \wedge 1}$ are homeomorphic, where $f \wedge 1:X \wedge Z \to Y \wedge Z$. Using this construction, new non-trivial examples of cohomology theories may be obtained.

(3) Given any theory h, we define theories $\bar{h}$ and $\bar{\bar{h}}$ by

$$\bar{h}^i = \bigoplus_{n \in Z} h^n \quad \text{(direct sum)}$$

$$\bar{h}^i = \prod_{n \in Z} h^n \quad \text{(direct product)}$$

for all i. Since direct sums and products preserve exactness, one verifies easily that this is a cohomology theory.

(4) We sketch now an important example of a cohomology theory: stable cohomotopy theory. Its origin is a theorem of K. Borsuk, which asserts that if X is a space of dimension $\leq 2n - 2$, the set $[X, S^n]$ of all homotopy classes of maps $X \to S^n$ has an abelian group structure; furthermore, one shows that the group $[X, S^n]$ is isomorphic to all the groups $[\Sigma^q X, S^{n+q}]$, $q \geq 0$, the isomorphisms being given by the suspension map $[f] \to [\Sigma f]$.

Consider a finite-dimensional CW-complex X and an integer k. Consider (for $k \geq 0$) the sequence

$$[X, S^k] \xrightarrow{\Sigma} [\Sigma X, S^{k+1}] \xrightarrow{\Sigma} \dots \longrightarrow [\Sigma^q X, S^{k+q}] \longrightarrow \dots$$

or, for $k < 0$, the sequence

$$[\Sigma^{-k} X, S^0] \longrightarrow (\Sigma^{-k+1} X, S^1] \to \dots \longrightarrow [\Sigma^{-k+q} X, S^q] \longrightarrow \dots$$

In either case, these sequences become sequences of abelian groups and homomorphisms and they stabilize; we write $\Pi^k(X)$ for the stable value. The family $\{\Pi^k; k \in Z\}$ defines a cohomology theory. Axioms (1. 1) and (1. 2) are trivially verified; as for axiom (1. 3), it is enough to observe that given a sequence $X \xrightarrow{f} Y \xrightarrow{i} C_f$ and a space Z, the homotopy extension property shows the exactness of the sequence of sets $[C_f, Z] \to [Y, Z] \to [X, Z]$. Exactness can be carried to the stable range since the direct limit preserves exactness* (Eilenberg, S. and Steenrod, N. : Foundations of Algebraic Topology, Princeton U. Press, Chap. VIII, Theorem 5. 4).

---

* This argument would only be needed if we allowed infinite-dimensional complexes X. For an N-dimensional complex X, $\pi^k(X) = [\Sigma^m X, S^{m+k}]$ for any $m \geq N - 2k + 2$.

(5) From stable cohomotopy we get generalized cohomotopy by defining, for a fixed space Y, the groups $\Pi^k_Y(X) = \varinjlim_n [\Sigma^n X, \Sigma^{n+k} Y]$.

(6) Real and complex K-theory are cohomology theories. These will be discussed later.

The coefficients of a theory $h = \{h^n; n \in Z\}$ are defined as the graded group $\check{h} = \{\check{h}^n; n \in Z\}$, where $\check{h}^n = h^n(S^0)$.

For example, the coefficients of stable cohomotopy are

$$(1.6) \qquad \Pi^n(S^0) = \begin{cases} (0) \text{ if } n > 0 \\ Z \text{ if } n = 0 \\ \text{stable } (-n)\text{-stem for } n < 0 . \end{cases}$$

Observe that the stable (-n)-stems, for $n < 0$, are finite groups by a classical result of Serre. We will make considerable use of this observation later.

Let h be a cohomology theory, $\{X_i\}_{1 \le i \le k}$ a finite family of spaces of $\mathscr{C}$.

(1.7) **Theorem.** For all $n \in Z$ we have,

$$h^n(X_1 \vee \ldots \vee X_k) = \bigoplus_{i=1}^{k} h^n(X_i) .$$

Proof. It suffices to prove the theorem for $k = 2$. Let $i_1 : X_1 \to X_1 \vee X_2$, $i_2 : X_2 \to X_1 \vee X_2$ be the inclusions into the wedge, $j : X_1 \vee X_2 \to X_2$, the retraction onto X_2, so that $ji_2 = 1$. We have the homotopy equivalence $C_{i_1} \simeq X_2$ and $X_1 \xrightarrow{i_1} X_1 \vee X_2 \xrightarrow{j} X_2 \simeq C_{i_1}$ gives rise to an exact sequence

$$\ldots \xleftarrow{\delta} h^n(X_1) \xleftarrow{i_1^*} h^n(X_1 \vee X_2) \underset{j^*}{\overset{i_2^*}{\rightleftarrows}} h^n(X_2) \longleftarrow \ldots$$

which splits since $i_2^* \circ j^*$ is the identity homomorphism. This implies that j^* is a monomorphism; exactness then gives us $\text{Im } \delta = \ker j^* = 0$ that is to say, $\delta = 0$ and i_1^* is an epimorphism.

Hence for each integer n we have a split exact sequence

$$0 \leftarrow h^n(X_1) \xleftarrow{i_1^*} h^n(X_1 \vee X_2) \underset{j^*}{\overset{i_2^*}{\rightleftarrows}} h^n(X_2) \leftarrow 0$$

which proves the statement for $k = 2$.

Theorem 1. 7 cannot in general be extended to the wedge of an arbitrary family of spaces $\{X_\alpha\}_\alpha$. There are theories h for which it is not true that

$$(1.8) \qquad h(\vee_\alpha X_\alpha) \cong \Pi_\alpha h(X_\alpha) .$$

A counter-example may be obtained by considering the theory $\bar{H}$ defined, for each k, by $\bar{H}^k = \bigoplus_{n \in Z} H^n$, where H is ordinary cohomology. Consider the family of spaces $X_i = S^i$, $i \geq 1$. We have

$$\bar{H}^k(\vee_i S^i) = \bigoplus_{n \in Z} H^n(\vee_i S^i) \cong \bigoplus_{n \geq 1} Z, \text{ while}$$

$$\Pi_{i \geq 1} \bar{H}^k(S^i) = \Pi_{i \geq 1} \left(\bigoplus_{n \in Z} H^n(S^i)\right) \cong \Pi_{i \geq 1} Z .$$

Remark. The isomorphism (1. 8) is a sufficient condition for the realizability of the theory h in the following sense: a cohomology theory $h = \{h^n; n \in Z\}$ is said to be realizable if for each $n \in Z$ there exists a space Y_n (possibly in a larger category than $\mathscr{C}$) such that the contravariant functors h^n and $[\quad, Y_n]$ are natura equivalent (Brown, E. H., Jr. : Abstract Homotopy Theory, Trans. Amer. Math. Soc. 119 (1965); 79-85). We now discuss realizable theories more fully.

Let $\mathscr{C}'$ be the category of all CW-complexes with base-point and the appropriate morphisms; clearly $\mathscr{C} \subseteq \mathscr{C}'$. A spectrum $\underline{Y}$ is a sequence of objects $Y_n \in \mathscr{C}'$, $n \in Z$, together with morphisms $g_n : \Sigma Y_n \rightarrow Y_{n+1}$.

It is possible to construct cohomology theories using spectra. Given a space $X \in \mathscr{C}$ and an integer k, consider the diagram

$$\begin{array}{ccc} [X, Y_k] & \xrightarrow{\pi_{(0)}} & [\Sigma X, Y_{k+1}] \\ & \Sigma \searrow \quad \nearrow g_k^* & \\ & [\Sigma X, \Sigma Y_k] & \end{array}$$

where $\pi_{(0)} = g_k^* \circ \Sigma$. Then for each positive integer n, we can define $\pi_{(n)}:[\Sigma^n X, Y_{k+n}] \to [\Sigma^{n+1} X, Y_{k+n+1}]$; actually, $\pi_{(n)}$ is a homomorphism of abelian groups if $n \geq 2$. Thus to each space X and integer k we associate an abelian group

$$h^k_{\underline{Y}}(X) = \varinjlim_n [\Sigma^n X, Y_{k+n}] \quad .$$

The natural transformations $\sigma^k : h^k_{\underline{Y}} \to h^{k+1}_{\underline{Y}} \circ \Sigma$ are easily defined by passing to the direct limit. It may be shown that σ^k thus defined is actually a natural equivalence (Whitehead, G. W.: 'Generalized Homology Theories', Trans. Amer. Math. Soc. 102 (1962); 227-283).

Since direct limits preserve exactness, we have a cohomology theory, $h_{\underline{Y}}$.

An important example of a spectrum is given by $Y_n = S^n$ for $n \geq 0$, $Y_n = \{\text{point}\}$ if $n < 0$, where $g_n : \Sigma S^n \to S^{n+1}$ is the natural homeomorphism for $n \geq 0$. $\underline{S} = \{S^n; g_n\}$ is called the Sphere Spectrum; the corresponding cohomology theory is just stable cohomotopy.

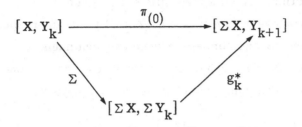

Let us look for a moment at a spectrum from the dual standpoint. Consider a space A with base-point a_0. A loop ℓ on A is a map $\ell : I, \dot{I} \to A, a_0$. The set $\Omega(A)$ of loops on A is topologized as a subspace of the function space A^I with the compact-open topology. There is an important natural one-to-one correspondence between maps $h: B \times I \to A$ and $h': B \to A^I$ which matches up continuous functions: given a map $h: B \times I \to A$, define $h'(b)(t) = h(b, t)$, for every $b \in B$, $t \in I$. Actually this correspondence induces a one-to-one correspondence between maps $\Sigma B \to A$ and maps $B \to \Omega A$, which in turn induces a one-to-one correspondence between homotopy classes of such maps,

$$(1.9) \qquad [\Sigma B, A] \longleftrightarrow [B, \Omega A] .$$

In fact, each side of (1.9) has a natural group structure and (1.9) is an isomorphism of groups. Thus, an equivalent definition of spectrum can be given as follows: $\underline{Y}$ is a family $\{Y_n, g_n'\}$ of spaces Y_n and maps $g_n': Y_n \to \Omega Y_{n+1}$. Given such a spectrum $\underline{Y}$, $X \in \mathscr{C}$ and $k \in Z$, we take the sequence

$$[X, Y_k] \xrightarrow{g_k'^*} [X, \Omega Y_{k+1}] \xrightarrow{\Omega g_{k+1}'^*} [X, \Omega^2 Y_{k+2}] \longrightarrow \cdots$$

and define $h^k_{\underline{Y}}(X) = \varinjlim_n [X, \Omega^n Y_{k+n}]$.

We say that $\underline{Y}$ is an <u>Ω-spectrum</u> if g_n' is a homotopy equivalence for every n. If so, we have $h^k_{\underline{Y}}(X) = [X, Y_k]$ and the cohomology theory is <u>representable.</u>

Consider the following examples of Ω-spectra.

Let G be a group and n an integer > 0. An <u>Eilenberg-MacLane</u> space of type (G, n) is a space $K(G, n)$ such that $\Pi_i(K(G, n)) = G$ if $i = n$ and zero otherwise. If G is a (countable) group, such a space can be realized as a (countable) CW-complex and, for a (countable) CW-complex L, ΩL has again the homotopy type of a (countable) CW-complex. Now one verifies that $\Omega K(G, n+1) \simeq K(G, n)$. The <u>Eilenberg-Mac Lane spectrum</u> (corresponding to the group G) is defined by taking $Y_n = K(G, n)$ if

$n \geq 1$, $Y_0 = G$ as a discrete space and $Y_n = \{\text{point}\}$ if $n < 0$, $g'_n : Y_n \xrightarrow{\simeq} \Omega Y_{n+1}$ is the natural homotopy equivalence for $n \geq 0$, g'_{-1} is the identity. The associated cohomology theory is ordinary cohomology.

Other examples are the so-called unitary and orthogonal spectra. Let $U(n)$ be the group of all unitary $(n \times n)$-matrices over the complex field C. Imbed $U(n)$ into $U(n+1)$ by sending the matrix $\sigma \in U(n)$ into $\begin{pmatrix} \sigma & 0 \\ 0 & 1 \end{pmatrix}$ and define $U = \bigcup_{n \geq 1} U(n)$ with the weak topology. The Bott Periodicity Theory (Bott, R. : The stable homotopy of the classical groups', Annals of Math. 70 (1959); 313-337) asserts that $\Omega^2 U \simeq U$. We then construct an Ω-spectrum by taking $Y_n = \Omega U$ if n is even and $Y_n = U$ if n is odd; g'_n is the identity if n is even and is the Bott map $U \to \Omega^2 U$ if n is odd. This spectrum defines a cohomology theory called complex K-Theory.

There is a real analogue based on the Bott Periodicity $\Omega^8 O \simeq O$, where O is the infinite orthogonal group. Here we take the spaces Y_n to be:

$$Y_n = O \text{ if } n \equiv -1 \pmod 8$$
$$Y_n = \Omega^k O \text{ if } n \equiv 7 - k \pmod 8), \quad 0 \leq k \leq 7.$$

The maps g'_n are taken to be the identity if $n \not\equiv -1 \pmod 8$ and the Bott map $O \to \Omega^8 O$ if $n \equiv -1 \pmod 8$. This spectrum gives rise to real K-Theory or KO-Theory. There are, of course, many other representable theories; prominent among these are the various cobordism theories.

We conclude this chapter by observing that if we wish to pass from reduced cohomology to non-reduced (free) cohomology we simply define, for any arbitrary space X, the space $X^+ = X \cup \{\text{point}\}$ (where the extra point will be taken as base-point), and set

$$(1.10) \qquad h^q_{\text{free}}(X) = h^q(X^+) .$$

Non-reduced cohomology is sometimes called <u>free cohomology</u> because if h is given by an Ω-spectrum $\underline{Y} = \{Y_q\}$ so that, for every $X \in \mathscr{C}$, $h^q(X) = [X, Y_q]$, the set of homotopy classes of <u>based</u> maps, then $h^q_{\text{free}}(X) = [X, Y_q]_f$, the set of <u>free</u> homotopy classes. We may also pass in the other direction as follows. Let (X, x_0) be a based space and let k be a free theory. The map $x_0 \to X$ induces $k(X) \to k(x_0)$ and we define a reduced theory $\tilde{k}$ by setting $\tilde{k}(X)$ equal to the kernel of this homomorphism. In fact, of course, the sequence

$$\tilde{k}(X) \to k(X) \to k(x_0)$$

splits naturally, so that

$$(1.11) \qquad k(X) = \tilde{k}(X) \oplus k(x_0) .$$

Notice that $k(x_0)$ is the coefficient group (for the theories k or $\tilde{k}$) the projection $k(X) \to k(x_0)$ is sometimes called the <u>augmentation.</u> The reader should verify that $(\tilde{k})_{\text{free}}$ is naturally equivalent to k and $(\tilde{h}_{\text{free}})$ is naturally equivalent to h.

2. Exact Couples and Spectral Sequences

Throughout this chapter we shall be dealing with the category **Ab** of abelian groups and homomorphisms, though our results would apply to a more general abelian category.

An abelian group E together with an endomorphism d such that $d^2 = 0$ is called a differential group. Let $Z = \ker d$, the subgroup of cycles of E, $B = \operatorname{im} d$, the subgroup of boundaries of E; B is clearly a subgroup of Z and the quotient $H(E) = Z/B$ is called the homology group of (E, d).

A spectral sequence is a sequence of differential groups $\{E_n, d_n; n \in \mathbf{Z}_+\}$ such that $E_{n+1} = H(E_n)$, $n = 0, 1, \ldots$.

A homomorphism of a spectral sequence $\{E_n, d_n; n \in \mathbf{Z}_+\}$ into a spectral sequence $\{E'_n, d'_n; n \in \mathbf{Z}_+\}$ is a sequence of homomorphisms $f_n: E_n \to E'_n$ such that 1) for all n, $d'_n f_n = f_n d_n$; 2) f_{n+1} is the map induced by f_n on homology, for all n.

One method of obtaining spectral sequences is given by the consideration of exact couples, introduced by Massey.

An exact couple is a pair of abelian groups D, E together with homomorphisms α, β and γ such that the triangle

(2.1)

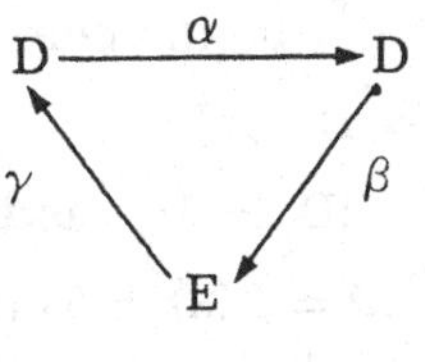

is exact. We denote the exact couple by $\mathfrak{C} = (D, E; \alpha, \beta, \gamma)$. Observe that the homomorphism $d = \beta\gamma: E \to E$ is such that $d^2 = \beta\gamma\beta\gamma = 0$, since $\gamma\beta = 0$ in view of exactness. Let $E_1 = H(E, d)$. It is easy to verify that $Z(E) = \gamma^{-1}(\alpha D)$ and

$B(E) = \beta(\ker \alpha)$, so

(2. 2) $$E_1 = \gamma^{-1}(\alpha D)/\beta(\ker \alpha).$$

Next we set $D_1 = \alpha D$ and define $\alpha_1 : D_1 \to D_1$ as the restriction of α to D_1. Furthermore we define the homomorphism $\beta_1 : D_1 \to E_1$ by $\alpha x \to \beta x + \beta(\ker \alpha)$, $x \in D$, and $\gamma_1 : E_1 \to D_1$ by $y + \beta(\ker \alpha) \to \gamma(y)$, $y \in E$. Some diagram chasing shows that β_1, γ_1 are well-defined and

(2. 3)

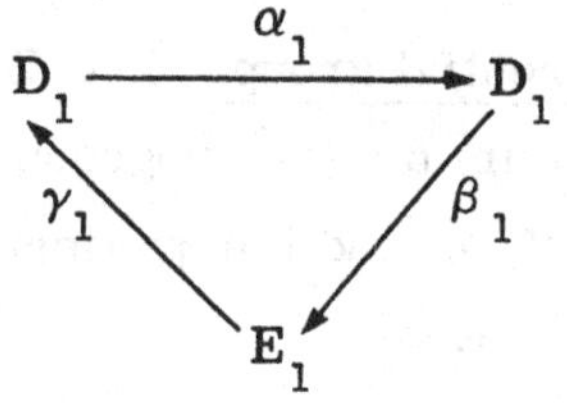

is an exact triangle. The exact couple $\mathfrak{C}' = (D_1, E_1; \alpha_1, \beta_1, \gamma_1)$ is called the derived couple of $\mathfrak{C}$.

The successive derived couples furnish the groups $E_2, E_3, \ldots$ such that $E_{n+1} = H(E_n, d_n)$, which gives a spectral sequence. Thus we have defined, in effect, a functor from exact couples to spectral sequences.

It is possible to give a direct description of the groups E_r without carrying out the successive derivations. Indeed we may describe the rth derived couple directly in terms of the original couple. For any integer $r \geq 0$, let α^r be the composite of α with itself r times. Define $D_r = \alpha^r D$ and set $E_r = \gamma^{-1}(\alpha^r D)/\beta(\ker \alpha^r)$. Define $\alpha_r : D_r \to D_r$ as the restriction of α to D_r, $\beta_r : D_r \to E_r$ by $\beta_r(\alpha^r x) = \beta(x) + \beta(\ker \alpha^r)$; notice that β_r is well defined, for if $\alpha^r x = \alpha^r x'$, then $x - x' \in \ker \alpha^r$ and $\beta(x) - \beta(x') = \beta(x - x') \in \beta(\ker \alpha^r)$. Define $\gamma_r : E_r \to D_r$ by setting $\gamma_r(y + \beta(\ker \alpha^r)) = \gamma(y) \in \alpha^r D$. In this way we get an exact couple

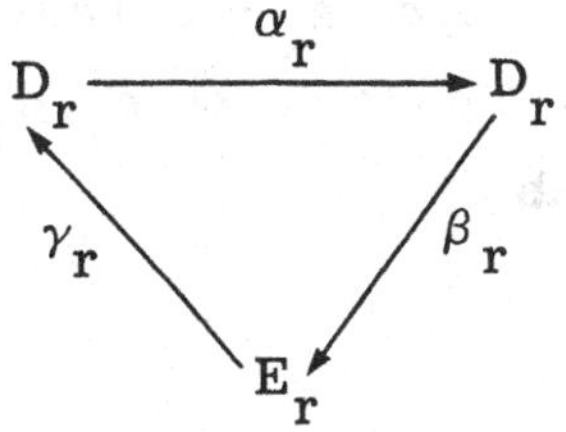

which is called the r-th derived couple $\mathfrak{C}_r$ of the exact couple $\mathfrak{C}$. Take $\mathfrak{C}_0 = \mathfrak{C}$ and $\mathfrak{C}_1 = \mathfrak{C}'$; one verifies that $\mathfrak{C}_{r+1} = \mathfrak{C}'_r$ (see MacLane, S. - Homology, Ch. XI).

A homomorphism of the exact couple $(D, E; \alpha, \beta, \gamma)$ into the exact couple $(D', E'; \alpha', \beta', \gamma')$ is a pair of homomorphisms $(f, g), f: D \to D'$, $g: E \to E'$ such that each square of the diagram below is commutative:

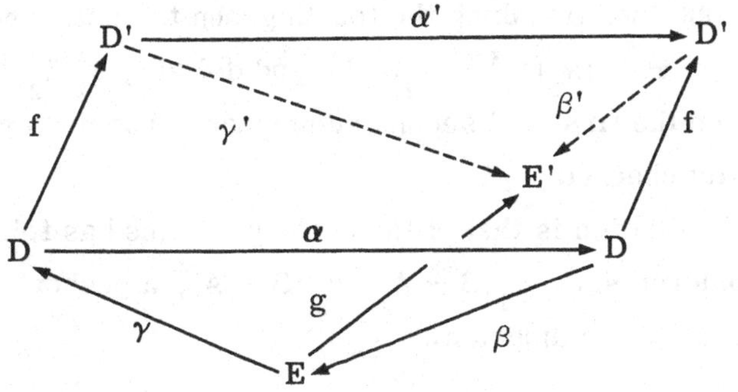

As one easily verifies, g commutes with the differential operators $d = \beta\gamma$ and $d' = \beta'\gamma'$ of the respective couples, so it passes to homology, $g_1: E_1 \to E'_1$; together with the restriction f_1 of f to a map of D_1 into D'_1, it defines a homomorphism (f_1, g_1) of the respective derived couples; and hence also a homomorphism of the respective spectral sequences.

Associated to each spectral sequence there is an important limit group E_∞. To get a clearer view of how such a group fits in with the spectral sequence obtained from an exact couple, we now consider a different approach to spectral sequences, due to Eckmann and Hilton.

Let $A_1, A_2, B \in \mathbf{Ab}$ and $\alpha_1 : A_1 \to B$, $\alpha_2 : A_2 \to B$ be two homomorphisms with the same codomain; a <u>pullback</u> of α_1, α_2 is a commutative diagram, where $R \in \mathbf{Ab}$,

$$(2.4) \qquad \begin{array}{ccc} R & \xrightarrow{i_1} & A_1 \\ {\scriptstyle i_2}\downarrow & & \downarrow{\scriptstyle \alpha_1} \\ A_2 & \xrightarrow{\alpha_2} & B \end{array}$$

with the following <u>universal property</u>: if $i'_1 : R' \to A_1$, $i'_2 : R \to A_2$ are such that $\alpha_1 i'_1 = \alpha_2 i'_2$ then there exists a <u>unique</u> homomorphism $h : R' \to R$ such that $i_1 h = i'_1$ and $i_2 h = i'_2$. It is easy to show that the pullback is unique up to canonical isomorphism. A pullback of α_1, α_2 may be obtained by taking the fourth group to be the group $R = \{(a_1, a_2) \in A_1 \oplus A_2 ; \alpha_1(a_1) = \alpha_2(a_2)\}$ and defining i_1, i_2 as the projections on the first and second summands. The universal property is easily checked.

The dual situation is that of the <u>pushout</u>, defined as follows: given two homomorphisms $\beta_1 : B \to A_1$, $\beta_2 : B \to A_2$, a pushout of β_1, β_2 is a commutative diagram

$$(2.5) \qquad \begin{array}{ccc} B & \xrightarrow{\beta_1} & A_1 \\ {\scriptstyle \beta_2}\downarrow & & \downarrow{\scriptstyle j_1} \\ A_2 & \xrightarrow{j_2} & K \end{array}$$

with the following universal property: if $j'_1 : A_1 \to K$, $j'_2 : A_2 \to K$ are such that $j'_1 \beta_1 = j'_2 \beta_2$, then there exists a <u>unique</u> homomorphism $k : K \to K'$ such that $kj_1 = j'_1$ and $kj_2 = j'_2$. Again, two groups K and K' which complete the diagram with this universal property are isomorphic. One such completion may be obtained by taking

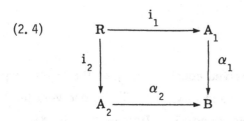

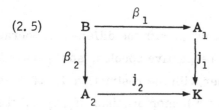

$K = (A_1 \oplus A_2)/C$, where C is the subgroup of $A_1 \oplus A_2$ consisting of all pairs $(\beta_1(b), -\beta_2(b))$, $b \in B$. The maps $j_1 : A_1 \to K$ and $j_2 : A_2 \to K$ are defined $j_1(a_1) =$ class of $(a_1, 0)$, denoted by $\overline{(a_1, 0)}$; likewise, $j_2(a_2) = \overline{(0, a_2)}$.

(2.6) **Lemma.** (i) <u>In a pullback of</u> $\alpha_1 : A_1 \to B$ <u>and</u> $\alpha_2 : A_2 \to B$, <u>if</u> α_1 <u>is a monomorphism, so is</u> i_2.

(ii) <u>In a pushout of</u> $\beta_1 : B \to A_1$ <u>and</u> $\beta_2 : B \to A_2$, <u>if</u> β_1 <u>is an epimorphism, so is</u> j_2.

Proof (see diagrams 2.4 and 2.5)

(i) Assume the pullback diagram is completed with the group $R = \{(a_1, a_2) \in A_1 \oplus A_2 ; \alpha_1(a_1) = \alpha_2(a_2)\}$, let $i_2(a_1, a_2) = i_2(a_1', a_2')$. Since $\alpha_1 i_1 = \alpha_2 i_2$, we get $\alpha_1 i_1(a_1, a_2) = \alpha_2 i_2(a_1', a_2')$ or $\alpha_1(a_1) = \alpha_1(a_1')$, which gives $a_1 = a_1'$ since α_1 is monic. Plainly, $i_2(a_1, a_2) = i_2(a_1', a_2')$ implies $a_2 = a_2'$. Hence i_2 is monic.

(ii) Let $\overline{(a_1, a_2)}$ be an arbitrary element of $K = A_1 \oplus A_2/C$. Since β_1 is an epimorphism, there is a $b \in B$ such that $\beta_1(b) = a_1$. Consider the element $a_2 + \beta_2(b) \in A_2$; by definition we have $j_2(a_2 + \beta_2(b)) = \overline{(0, a_2 + \beta_2(b))}$. But $(a_1, a_2) - (0, a_2 + \beta_2(b)) = (\beta_1(b), -\beta_2(b)) \in C$. Thus $j_2(a_2 + \beta_2(b)) = \overline{(a_1, a_2)}$. Notice that, in fact, if α_1 is regarded as an inclusion, R is nothing but $\alpha_2^{-1}(A_1)$; and if β_1 projects B onto $A_1 = B/L$, then j_2 projects A_2 onto $K = A_2/\beta_2 L$.

Consider now the exact couple $(D, E; \alpha, \beta, \gamma)$. The homomorphism $\alpha : D \to D$ can be factored as a composite of an epimorphism followed by a monomorphism $\alpha : D \xrightarrow{\sigma} D_1 \xrightarrow{\rho} D$, where $D_1 = \alpha D$.

Take the pullback of ρ and γ:

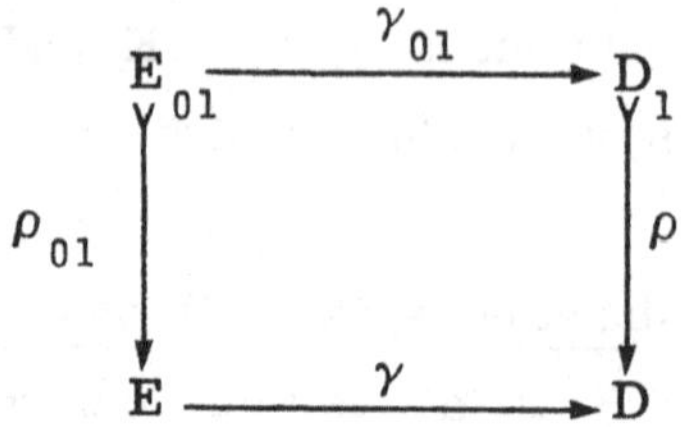

where we have taken $E_{01} = \{(d_1, e) \in D_1 \oplus E;\ \rho(d_1) = \gamma(e)\}$, and γ_{01}, ρ_{01} are the projections on the first and second factors respectively. Note that ρ_{01} is monic because ρ is monic. We shall show that E_{01} picks up all cycles of $d = \beta\gamma$; more precisely;

(2.7) **Lemma.** Given the following diagram of objects and morphisms

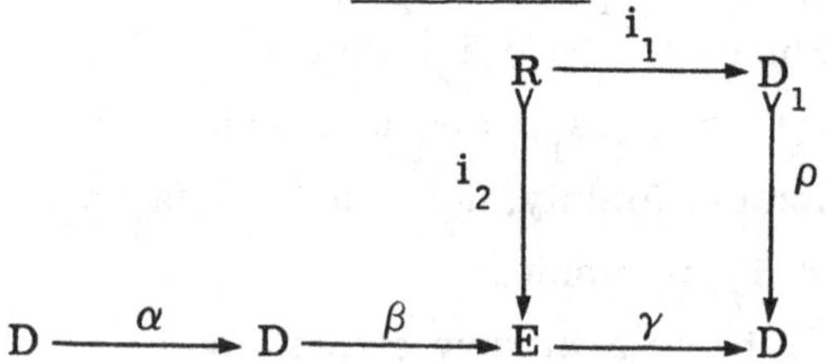

such that 1) the long horizontal sequence is exact; 2) the square is a pullback of ρ and γ, with ρ the monomorphic component of α and $D_1 = \alpha D$. Then $R \cong \operatorname{im} i_2 = \ker(\beta\gamma)$.

Proof. Assume $R = \{(d_1, e) \in D_1 \oplus E;\ \rho(d_1) = \gamma(e)\}$. If $e \in \operatorname{im} i_2$ $e = i_2(d_1, e)$ with $\rho(d_1) = \gamma(e)$, then $\beta\gamma(e) = \beta\rho(d_1) = \beta\alpha(d') = 0$, because $d_1 = \alpha(d')$ for a certain $d' \in D$. Conversely, given $e \in \ker(\beta\gamma)$, $\gamma(e) \in \ker\beta = \operatorname{im}\alpha$; thus $\gamma(e) = \alpha(d')$ with $d' \in D$ and $e = i_2(\alpha(d'), e)$. (This lemma becomes obvious if we take $R = \gamma^{-1}(D_1)$.)

Now define a homomorphism $\beta_{01}: D \to E_{01}$ which takes any element $d' \in D$ into the pair $(0, \beta(d')) \in E_{01}$, and consider the pushout of σ, β_{01} given by:

$$\begin{array}{ccc} D & \xrightarrow{\quad\sigma\quad} & D_1 \\ {\scriptstyle \beta_{01}}\downarrow & & \downarrow{\scriptstyle \beta_{11}} \\ E_{01} & \xrightarrow{\quad\sigma_{01}\quad} & E_{11} \end{array}$$

Next we show that β_{01} takes the kernel of σ into the kernel of σ_{01}.

(2.8) **Lemma.** Given any pushout,

$$\begin{array}{ccc} B & \xrightarrow{\quad\beta_1\quad} & A_1 \\ {\scriptstyle \beta_2}\downarrow & & \downarrow{\scriptstyle j_1} \\ A_2 & \xrightarrow{\quad j_2\quad} & K \end{array}$$

of β_1, β_2, then $\beta_2(\ker \beta_1) = \ker j_2$.

Proof. Take $K = (A_1 \oplus A_2)/C$, where $C = \{(\beta_1(b), -\beta_2(b));\ b \in B\}$. For any $a_2 \in \ker j_2$, $j_2(a_2) = \overline{(0, a_2)} = 0$, and thus $(0, a_2) \in C$. This means that there is $b \in B$ such that $\beta_1(b) = 0$ and $a_2 = \beta_2(-b)$; and thus that $a_2 \in \beta_2(\ker \beta_1)$.

Conversely, if $a_2 \in \beta_2(\ker \beta_1)$, $a_2 = \beta_2(b)$ with $\beta_1(b) = 0$ and thus $j_2(a_2) = j_2\beta_2(b) = j_1(\beta_1(b)) = 0$. Hence $a_2 \in \ker j_2$.

Going back to the definition of the group E_{11}, we get: $E_{11} \cong E_{01}/\ker \sigma_{01} \cong E_{01}/\beta_{01}(\ker \sigma) \cong \ker(\beta\gamma)/\beta_{01}(\ker \sigma)$, using (2.7) and (2.8). On the other hand, $\beta_{01}(\ker \sigma) = \beta_{01}(\ker \alpha) = \beta_{01}(\operatorname{im} \gamma) \cong \beta(\operatorname{im} \gamma)$, the isomorphism being given by ρ_{01}. Thus

$$E_{11} \cong \ker(\beta\gamma)/\operatorname{im}(\beta\gamma) ,$$

that is, E_{11} is the homology group of $(E, d = \beta\gamma)$.

This construction of E_{11} is better seen in the diagram

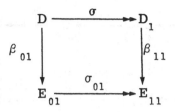

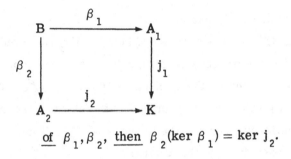

(2.9)
$$
\begin{array}{ccccccccc}
D_1 & \xrightarrow{\alpha_1} & D_1 & \xrightarrow{\beta_{11}} & E_{11} & \xrightarrow{\gamma_{11}} & D_1 & \xrightarrow{\alpha_1} & D_1 \\
\sigma\uparrow & & \sigma\uparrow & \text{push out} & \uparrow\sigma_{01} & & \| & & \| \\
D & \xrightarrow{\alpha} & D & \xrightarrow{\beta_{01}} & E_{01} & \xrightarrow{\gamma_{01}} & D_1 & \xrightarrow{\alpha} & D_1 \\
\| & & \| & & \downarrow\rho_{01} & \text{pull back} & \downarrow & & \downarrow\rho \\
D & \xrightarrow{\alpha} & D & \xrightarrow{\beta} & E & \xrightarrow{\gamma} & D & \xrightarrow{\alpha} & D
\end{array}
$$

where each horizontal line is exact.

If we write Q^ρ for the process of getting E_{01} as a pullback of ρ, γ and Q_σ for the process of getting E_{11} as a pushout of σ, β_{01}, then E_{11} is obtained by the composite process $Q_\sigma Q^\rho$. We might inquire about getting E_{11} by inverting the steps: first the pushout then the pullback, that is, the process $Q^\rho Q_\sigma$. That situation would be portrayed by the diagram

$$
\begin{array}{ccccccccc}
D_1 & \xrightarrow{\alpha_1} & D_1 & \xrightarrow{\beta_{11}} & \overline{E}_{11} & \xrightarrow{\gamma_{11}} & D_1 & \xrightarrow{\alpha_1} & D_1 \\
\| & & \| & & \downarrow\rho_{10} & \text{pull back} & \downarrow\rho & & \downarrow\rho \\
D_1 & \xrightarrow{\alpha_1} & D_1 & \xrightarrow{\beta_{10}} & E_{10} & \xrightarrow{\gamma_{10}} & D & \xrightarrow{\alpha} & D \\
\sigma\uparrow & & \sigma\uparrow & \text{push out} & \uparrow\sigma_{10} & & \| & & \| \\
D & \xrightarrow{\alpha} & D & \xrightarrow{\beta} & E & \xrightarrow{\gamma} & D & \xrightarrow{\alpha} & D
\end{array}
$$

where $E_{10} = (D_1 \oplus E)/C$ and $C = \{(\sigma d, -\beta d);\ d \in D\}$. The other maps are defined by $\beta_{01}(d_1) = \overline{(d_1, 0)}$ for all $d_1 \in D_1$, $\sigma_{01}(e) = \overline{(0, e)}$ for all $e \in E$ and $\gamma_{01}\overline{(d_1, e)} = \gamma(e)$, for every $\overline{(d_1, e)} \in E_{10}$. One then shows that the middle horizontal line is exact. Using (2.7), $\overline{E}_{11} \cong \operatorname{im} \rho_{10} \cong \ker(\beta_{10}\gamma_{10})$ and by (2.8), $\ker \sigma_{10} = \beta(\ker \sigma)$.

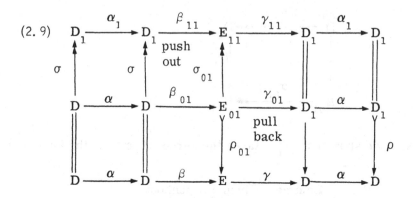

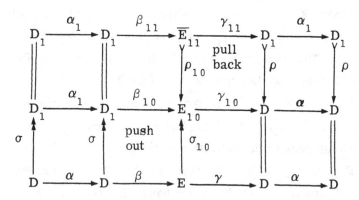

Hence $E_{10} \cong E/\beta(\ker\sigma) \cong E/\mathrm{im}\,\beta\gamma$. Notice that $\overline{E}_{11} \cong \ker(\beta_{10}\gamma_{10})$ is a subgroup of $E/\mathrm{im}\,\beta\gamma$. We now show that we have in fact $\overline{E}_{11} \cong \ker\beta\gamma/\mathrm{im}\,\beta\gamma$. For this, it suffices to show that σ_{10} takes $\ker\beta\gamma$ onto $\ker\beta_{10}\gamma_{10}$.

To prove the inclusion $\sigma_{10}(\ker\beta\gamma) \subset \ker\beta_{10}\gamma_{10}$ we prove that $\sigma_{10}(\ker\beta\gamma) \subset \mathrm{im}\,\rho_{10}$. In fact, if $e \in E$ is such that $\beta\gamma(e) = 0$ then $\gamma(e) \in \mathrm{im}\,\alpha$ and thus there is a $d \in D$ such that $\gamma(e) = \sigma(d)$. Take $(\sigma(d), \overline{(0, e)}) \in \overline{E}_{11}$; this element is such that $\rho_{10}(\sigma(d), \overline{(0, e)}) = \overline{(0, e)} = \sigma_{10}(e)$.

Now we show that σ_{10} maps onto $\ker(\beta_{10}\gamma_{10})$. In fact if $\overline{(d_1, e_1)} \in \ker\beta_{10}\gamma_{10}$, $\beta_{10}\gamma_{10}\,\overline{(d_1, e_1)} = \overline{(\gamma(e_1), 0)} = 0$ and thus there exists $d_2 \in D$ such that $\gamma(e_1) = \sigma(d_2)$ and $\beta(d_2) = 0$. On the other hand, $d_1 \in D_1$ means that $d_1 = \alpha(d_3)$ for some $d_3 \in D$. Now take $e = e_1 + \beta(d_3) \in \ker\beta\gamma$; then $\sigma_{10}(e) = \overline{(0, e_1 + \beta(d_3))} = \overline{(d_1, e_1)}$ since $(d_1, e_1) - (0, e_1 + \beta(d_3)) = (\sigma(d_3)) \in C$.

We have shown that $\overline{E}_{11} \cong E_{11}$, the homology group of $(E, d = \beta\gamma)$. Thus the processes $Q^{\rho} Q_{\sigma}$ and $Q_{\sigma} Q^{\rho}$ produce the same final result. In view of this, we denote the process of obtaining the homology of $(E, d = \beta\gamma)$, by means of a pullback and pushout (performed in either order) by Q^{ρ}_{σ}, and we have $Q^{\rho}_{\sigma} = Q^{\rho} Q_{\sigma} = Q_{\sigma} Q^{\rho}$.

Recall that $D_m = \alpha^m D$, where α^m is the composite of α with itself m times; also, α_m has been defined as the restriction of α to D_m. Factor α_m as an epimorphism followed by a monomorphism: $\alpha_m = \rho_m \sigma_m$, $\sigma_m : D_m \twoheadrightarrow D_{m+1}$ and $\rho_m : D_{m+1} \rightarrowtail D_m$. We should notice at this point that if $\nu_n = \rho\rho_1 \ldots \rho_{n-1}$ and $\eta_m = \sigma_{m-1}\sigma_{m-2} \ldots \sigma_1\sigma$ then

$$Q^{\nu_n} = Q^{\rho_{n-1}} Q^{\rho_{n-2}} \ldots Q^{\rho_1} Q^{\rho} \text{ and } Q_{\eta_m} = Q_{\sigma_{m-1}} Q_{\sigma_{m-2}} \ldots$$

$\ldots Q_{\sigma_1} Q_{\sigma}$. Now look in particular at the result of the process

$Q_{\eta_m} Q^{\nu_n}$, which corresponds to the diagram:

$$\begin{array}{ccccccccc}
D_m & \xrightarrow{\alpha_m} & D_m & \xrightarrow{\beta_{mn}} & E_{mn} & \xrightarrow{\gamma_{mn}} & D_n & \xrightarrow{\alpha_n} & D_n \\
\uparrow{\scriptstyle \eta_m} & & \uparrow{\scriptstyle \eta_m} & \text{push out} & \uparrow{\scriptstyle \sigma_{0n}} & & \| & & \| \\
D & \xrightarrow{\alpha} & D & \xrightarrow{\beta_{0n}} & E_{0n} & \xrightarrow{\gamma_{0n}} & D_n & \xrightarrow{\alpha_n} & D_n \\
\| & & \| & & \downarrow{\scriptstyle \rho_{0n}} & \text{pull back} & \downarrow{\scriptstyle \nu_n} & & \downarrow{\scriptstyle \nu_n} \\
D & \xrightarrow{\alpha} & D & \xrightarrow{\beta} & E & \xrightarrow{\gamma} & D & \xrightarrow{\alpha} & D
\end{array}$$

it shows that $E_{mn} \cong \mathrm{im}\, \rho_{0n}/\ker \sigma_{0n}$ and, applying (2. 8) we get $E_{mn} \cong \mathrm{im}\, \rho_{0n}/\beta_{0n}(\ker \eta_m) \cong \mathrm{im}\, \rho_{0n}/\beta(\ker \alpha^m)$. On the other hand, $\mathrm{im}\, \rho_{0n} = \gamma^{-1}(\alpha^n D)$. In fact, for every $e \in \mathrm{im}\, \rho_{0n}$ we have $e = \rho_{0n}(d_n, e)$ with $d_n \in D_n$, and furthermore, $\gamma(e) = \alpha_n(d_n) \in \alpha^n D$; in other words $e \in \gamma^{-1}(\alpha^n D)$. If $e \in \gamma^{-1}(\alpha^n D)$, $\gamma(e)$ is equal to a certain $d_n \in \alpha^n D$, so $(d_n, e) \in E_{0n}$ and $e = \rho_{0n}(d_n, e)$. This shows that

$$(2.10) \qquad E_{mn} \cong \gamma^{-1}(\alpha^n D)/\beta(\ker \alpha^m) .$$

In particular, if $m = n$, $E_{nn} \cong \gamma^{-1}(\alpha^n D)/\beta(\ker \alpha^n)$ is isomorphic to the term E_n in the n-th derived couple of $(D, E; \alpha, \beta, \gamma)$.

Observe that for each positive integer n, we can consider D_n either as the subgroup $\alpha^n D = \mathrm{im}\, \nu_n$ or D, or as the quotient group $D_n = \mathrm{im}\, \eta_n \cong D/\ker \alpha^n$; moreover, $\rho_n : D_{n+1} \rightarrowtail D_n$ and $\sigma_n : D_n \twoheadrightarrow D_{n+1}$. The system $\{D_n, \sigma_n; n \in \mathbf{Z}_+\}$ defines a direct limit

$$(2.11) \qquad U = \varinjlim_n D_n = D/\bigcup_n \ker \alpha^n .$$

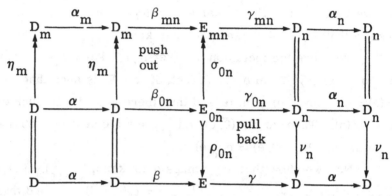

On the other hand, the system of groups and monomorphisms $\{D_n, \rho_n\}$ defines an inverse limit $I = \varinjlim_{n} D_n$. Precisely, the <u>inverse limit</u> of $\{D_n, \rho_n\}$ is the subgroup of the product $\prod_n D_n$ consisting of those elements $x = (x_n)$ such that $\rho_n(x_{n+1}) = x_n$. Because ρ_n is an inclusion, we have $I = \bigcap_n \alpha^n D$.

There is an obvious projection $\eta : D \twoheadrightarrow U$ and an embedding $\nu : I \rightarrowtail D$. By the previous argument, either one of the processes $Q_\eta \; Q^\nu$ or $Q^\nu \; Q_\eta$ defines a group E_∞ and a diagram

(2. 12)
$$\begin{array}{ccccccccc}
U & \xrightarrow{\alpha'} & U & \longrightarrow & E_\infty & \longrightarrow & I & \xrightarrow{\alpha''} & I \\
\uparrow{\scriptstyle \eta} & & \uparrow{\scriptstyle \eta} & & & & \downarrow{\scriptstyle \nu} & & \downarrow{\scriptstyle \nu} \\
D & \xrightarrow{\alpha} & D & \xrightarrow{\beta} & E & \xrightarrow{\eta} & D & \xrightarrow{\alpha} & D
\end{array}$$

with exact horizontal lines; whence, in particular, the exact sequence

(2. 13) $$0 \to \operatorname{coker} \alpha' \to E_\infty \to \ker \alpha'' \to 0 .$$

One should notice that E_∞ is obtained in (2. 12) in only two stages from the original couple; application of a previous argument shows that

(2. 14) $$E_\infty \cong \gamma^{-1}(\operatorname{im} \nu)/\beta(\ker \eta) .$$

Let us analyse more closely the pictures created by the definition of E_∞. Consider the diagram

$$
(2.15)\qquad
\begin{array}{ccccccccc}
D_m & \xrightarrow{\alpha_m} & D_m & \xrightarrow{\beta_{m0}} & E_{m0} & \xrightarrow{\gamma_{m0}} & D & \xrightarrow{\alpha} & D \\
\uparrow{\scriptstyle \eta_m} & & \uparrow{\scriptstyle \eta_m} & & & & \| & & \| \\
D & \xrightarrow{\alpha} & D & \xrightarrow{\beta} & E & \xrightarrow{\gamma} & D & \xrightarrow{\alpha} & D
\end{array}
$$

with exact rows. Keep the right hand side of (2.15) fixed and set $E_{m+1,0}$ as the pushout of σ_m, β_{m0}; this defines the following diagram

$$
\begin{array}{ccccccccc}
D_{m+1} & \xrightarrow{\alpha_{m+1}} & D_{m+1} & \longrightarrow & E_{m+1,0} & \longrightarrow & D & \xrightarrow{\alpha} & D \\
\uparrow & & \uparrow & & \uparrow & & \| & & \| \\
D_m & \xrightarrow{\alpha_m} & D_m & \xrightarrow{\beta_{m0}} & E_{m,0} & \longrightarrow & D & \xrightarrow{\alpha} & D
\end{array}
$$

Calling $E_{\infty,0} = \varinjlim E_{m,0}$ and considering that direct limits preserve exactness we obtain the exact sequence

$$
(2.16)\qquad U \to U \to E_{\infty,0} \to D \to D
$$

From (2.16), we define $E_{\infty,1}$ as the pullback of $\rho : D_1 \rightarrowtail D$ and of the map $E_{\infty,0} \to D$, obtaining the diagram,

$$
\begin{array}{ccccccccc}
U & \xrightarrow{\alpha'} & U & \longrightarrow & E_{\infty,1} & \longrightarrow & D_1 & \xrightarrow{\alpha_1} & D_1 \\
\| & & \| & & \downarrow{\scriptstyle \rho_0^1} & & \downarrow{\scriptstyle \rho} & & \downarrow{\scriptstyle \rho} \\
U & \xrightarrow{\alpha'} & U & \longrightarrow & E_{\infty,0} & \longrightarrow & D & \longrightarrow & D
\end{array}
$$

Proceeding in this way we obtain an inverse system of groups $\{E_{\infty,n}, \rho^n_{n-1}; n \in \mathbf{Z}_+\}$, where $\rho^n_{n-1} : E_{\infty,n} \rightarrowtail E_{\infty,n-1}$ are monomorphisms.

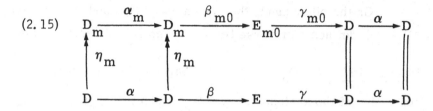

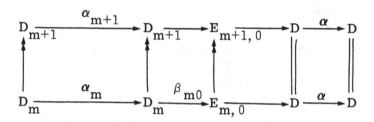

Though in general inverse limits do not preserve exactness, in the present case, due to the circumstance that the maps ρ are imbeddings, the inverse limits are intersections, namely, $\varprojlim E_{\infty,n} = \bigcap_n E_{\infty,n}$, $I = \varprojlim D_n = \bigcap_n \alpha^n D$. Hence we obtain a diagram

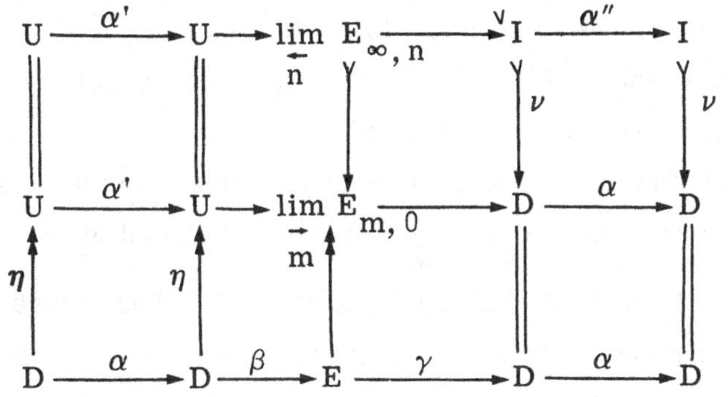

with exact rows. *

If we first consider the monomorphisms ρ_n and take pull-backs going towards the inverse limit and then pass to direct limits, we obtain the diagram

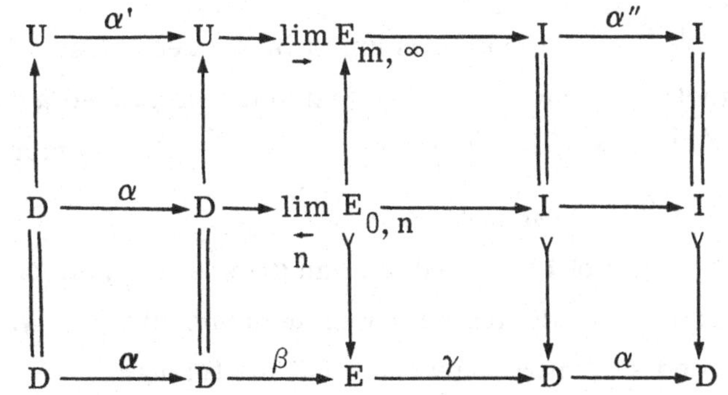

* A more careful scrutiny of the argument shows that we only require that direct limits preserve pushout and inverse limits preserve pullbacks. Thus our conclusions apply to a more general situation (see Eckmann, B. and Hilton, P. J., Exact couples in an abelian category, Journ. of Algebra, 1966, pp. 38-87).

Comparing the last two diagrams with (2.12) we obtain

$$(2.17)\qquad E_\infty \cong \varprojlim_n \left(\varinjlim_m E_{m,n}\right) \cong \varinjlim_m \left(\varprojlim_n E_{m,n}\right).$$

We now revert to (2.13). In applying spectral sequence techniques, it may well happen, under special circumstances, that one of coker α' or ker α'' is trivial; in such cases, E_∞ is isomorphic to the other. This is the situation where spectral sequences can give a lot of information.

Remark. The homomorphism $\alpha':U \to U$ is monic. In fact, α' which maps $U = D/\bigcup_n \ker \alpha^n \to D/\bigcup_n \ker \alpha^n$ is induced by α; now suppose that $x \in D$ is such that $\alpha(x) \in \bigcup_n \ker \alpha^n$. Then there exists n such that $\alpha(x) \in \ker \alpha^n$, that is $\alpha^n(\alpha(x)) = \alpha^{n+1}(x) = 0$. Hence $x \in \bigcup_n \ker \alpha^n$. One can give examples, however, to show that α'' is not necessarily epic. This shows that the monicity of α' does not 'dualize' in the categorical sense.

We can define exact couples and the associated spectral sequence when more structure is given. For example, define a <u>bi-graded abelian group</u> as a direct sum $D = \bigoplus_{p,q} D^{p,q}$ of abelian groups $D^{p,q}$. A homomorphism of bi-degree (r, s) of the bi-graded abelian group $D = \oplus D^{p,q}$ into the bi-graded abelian group $E = \oplus E^{p,q}$ is a direct sum $\alpha = \bigoplus_{p,q} \alpha^{p,q}$ of homomorphisms $\alpha^{p,q}:D^{p,q} \to E^{p+r,q+s}$, for each p, q.

An exact couple of bi-graded abelian groups is a pair of bi-graded abelian groups D, E together with homomorphisms α, β and γ of bi-graded abelian groups such that the triangle

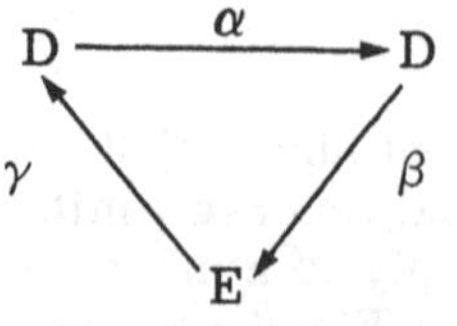

is exact.

In particular, if α is of bi-degree $(-1, 0)$, β of bi-degree $(1, 1)$ and γ of bi-degree $(0, 0)$, the homomorphism α_n, β_n and γ_n of the n-th derived couple will have bi-degrees $(-1, 0)$, $(n+1, 1)$ and $(0, 0)$ respectively. This can be shown by observing the definition of these homomorphisms. Hence the differential operator $d_n = \beta_n \gamma_n : E_n \to E_n$ will have bi-degree $(n+1, 1)$.

We shall find this situation in the spectral sequence we set up in Chapter III.

Remark. There is one technical point which arises in generalizing the theory of this chapter to graded (or bi-graded) couples. In the theory described we factorize α as $\rho\sigma : D \twoheadrightarrow D_1 \rightarrowtail D$. In the graded case, it is preferable to factorize α as $\rho\omega\sigma : D \xrightarrow{\sigma} D'_1 \xrightarrow{\omega} D''_1 \xrightarrow{\rho} D$, where ρ, σ have degree 0 and ω is an isomorphism carrying the degree of α. In this way all vertical morphisms in such diagrams as (2.9), (2.12), etc. have zero degree and the horizontal morphisms have the same degree as those on the lowest level. We have, however, to observe that the isomorphism $D'_n \overset{\omega_n}{\cong} D''_n$ will satisfy degree $\omega_n = n$ degree α; D'_n will appear in the left hand part of the diagram (replacing D_n) and D''_n in the right hand part (replacing D_n).

3. The Generalized Atiyah–Hirzebruch Spectral Sequence

Let h be a generalized cohomology theory defined on the category $\mathscr{C}$ of finite-dimensional cell-complexes with basepoint. If $X \in \mathscr{C}$, let X_p = p-dimensional skeleton of X, i.e. the subcomplex of X formed by all cells of dimension $\leq p$. If $p < 0$, let X_p be the basepoint. For each p we have an exact cohomology sequence

$$(3.1) \quad \ldots \longrightarrow h^q(X_p, X_{p-1}) \xrightarrow{\gamma^{p,q}} h^q(X_p) \xrightarrow{\alpha^{p,q}} h^q(X_{p-1}) \xrightarrow{\beta^{p-1,q}} h^{q+1}(X_p, X_{p-1}) \longrightarrow \ldots$$

Consider the bigraded groups obtained by setting $D = \oplus D^{p,q}$, where $D^{p,q} = h^q(X_p)$, and $E = \oplus E^{p,q}$, where $E^{p,q} = h^q(X_p, X_{p-1})$. Then the family of exact sequences above may be written as an exact couple of bigraded abelian groups,

(3.2)

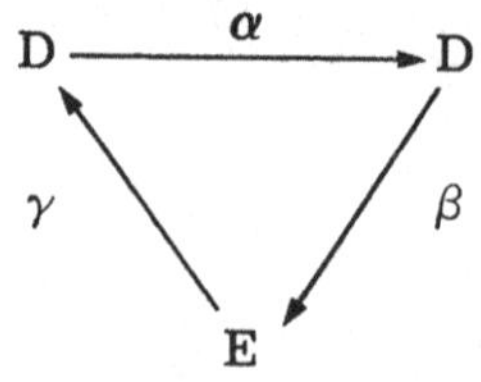

where α is of bidegree $(-1, 0)$, β of bidegree $(1, 1)$ and γ of bidegree $(0, 0)$. This exact couple will be denoted by $EC(X, h) = (D, E; \alpha, \beta, \gamma)$. Consider the spectral sequence associated with this couple. We saw in Chap. II that $E_n = \gamma^{-1}(\alpha^n D)/\beta\,(\ker \alpha^n)$ and that the differential $d_n : E_n \to E_n$, which is induced by $\beta\alpha^{-n}\gamma$ is of bidegree (n+1, 1).

We now identify the limit term E_∞. Consider the short exact sequence (2. 13), $0 \to \text{coker } \alpha' \to E_\infty \to \ker \alpha'' \to 0$, which is deduced from the limit sequence in (2. 12)

$$U \xrightarrow{\alpha'} U \longrightarrow E_\infty \longrightarrow I \xrightarrow{\alpha''} I .$$

The component $U^{p,q}$ of $U = D/\bigcup_n \ker \alpha^n$ is given by $U^{p,q} = D^{p,q}/(\bigcup_n \ker \alpha^n)^{p,q}$. To determine $\bigcup_n \ker \alpha^n$, observe that by iterating α enough times, $\ldots \to D^{p,q} \to D^{p-1,q} \to D^{p-2,q} \to \ldots$, one reaches a term $D^{p-k,q} = h^q(X_{p-k}) = h^q(\text{pt.}) = 0$ if $k > p$. So that $(\bigcup_n \ker \alpha^n)^{p,q} = E^{p,q} = 0$. Thus in this case the sequence (2. 13) shortens to

$$(3.\,3) \qquad 0 \longrightarrow E_\infty \xrightarrow[\approx]{\gamma''} \ker \alpha'' \longrightarrow 0 ,$$

γ'' being an isomorphism of bigraded groups since γ'' has bidegree (0, 0).

The group on the right side of this last isomorphism may be described as follows. If we take into consideration the graduation of the group $I = \bigcap_n \alpha^n D$, we see that

$$I^{p,q} = \bigcap_n \alpha^n D^{p+n,q}$$

Since dim X is finite, the iteration of the restriction map $\alpha^{p,q}: h^q(X_p) \to h^q(X_{p-1})$ a sufficient number of times gives us a stationary value $I^{p,q} = \alpha^n D^{p+n,q} = \text{Im}\,(h^q(X) \to h^q(X_p))$. We have the following canonical identifications:

$$I^{p,q} = \mathrm{im}(h^q(X) \to h^q(X_p)) \cong h^q(X)/\ker(h^q(X) \to h^q(X_p)) ,$$

$$I^{p-1,q} = \mathrm{im}(h^q(X) \to h^q(X_{p-1})) \cong h^q(X)/\ker(h^q(X) \to h^q(X_{p-1})) ,$$

and $\alpha'': I^{p,q} \to I^{p-1,q}$ is identified to the quotient map

$$h^q(X)/\ker(h^q(X) \to h^q(X_p)) \xrightarrow{\alpha''} h^q(X)/\ker(h^q(X) \to h^q(X_{p-1})) .$$

Consequently, we have

$$(3.4) \qquad (\ker \alpha'')^{p,q} = \frac{\ker(h^q(X) \to h^q(X_{p-1}))}{\ker(h^q(X) \to h^q(X_p))} .$$

To deduce from (3.4) a convenient description of $E_\infty^{p,q}$, we recall the following definitions.

A (descending) filtration of an abelian group A is a sequence F(A) of subgroups of A, $-\infty < p < +\infty$,

$$\ldots \subset F^{p+1}(A) \subseteq F^p(A) \subseteq F^{p-1}(A) \subseteq \ldots$$

The graded group $GA = GF(A)$ associated with this filtration is the graded group defined by $G^p(A) = F^p(A)/F^{p+1}(A)$.

Consider the filtration of $h^q(X)$ defined by $F^p(h^q(X)) = \ker(h^q(X) \to h^q(X_{p-1}))$. We have $F^{p+1}h^q(X) \subseteq F^p h^q(X)$, and if $k = \dim X$, the filtration $F^p h^q(X)$ is finite,

$$0 = F^{k+1}h^q(X) \subseteq F^k h^q(X) \subseteq \ldots \subseteq F^0 h^q(X) = h^q(X) .$$

In terms of the filtration above, we may write

(3.5) $$(\ker \alpha'')^{p,q} = F^p h^q(X) / F^{p+1} h^q(X) = G^p h^q(X) ,$$

or, more briefly, $\ker \alpha'' \cong Gh(X)$. The foregoing considerations may be summarized as follows.

(3.6) **Theorem.** The limit term of the spectral sequence associated with the exact couple $EC(X, h)$ is given by $E_\infty^{p,q} \cong G^p h^q(X)$. That is, E_∞ is isomorphic to the graded group associated with the filtration $F^p h^q(X) = \ker(h^q(X) \to h^q(X_{p-1}))$.

We will see later (Lemma (3.13)) that the spectral sequence converges finitely in the strong sense that if $\dim X = k$, then $E_k = E_{k+1} = \dots = E_\infty$, so that we reach the graded group associated with the filtration of $h(X)$ after a finite number of steps through the spectral sequence.

We now turn our attention to the early terms of the spectral sequence.

To analyse further the early terms of the spectral sequence associated with the exact couple $EC(X, h)$, we make an additional assumption about the cohomology theory h. By definition $E_0^{p,q} = h^q(X_p, X_{p-1}) = h^q(X_p/X_{p-1})$. On the other hand $X_p = X_{p-1} \cup \{e_\alpha^p\}$, so that $X_p/X_{p-1} = \bigvee_\alpha S_\alpha^p$.

We now make the following assumption on our cohomology theory h:

(3.7) $h^q(\bigvee_\alpha S_\alpha^p) \cong \prod_\alpha h^q(S_\alpha^p)$, for all p, q and index sets $\{\alpha\}$.

Recall that this property follows from the axioms if α ranges over a finite set. Thus no additional hypothesis on h is necessary if we confine attention to finite complexes. From now on, all theories will be assumed to satisfy (3. 7) unless otherwise stated.

Using (3. 7) and desuspending, we have

$$E_0^{p,q} \cong \prod_\alpha h^q(S^p_\alpha) \cong \prod_\alpha h^{q-p}(S^0),$$

where α runs over the p-cells of X.

Write $\check{h}^n = h^n(S^0) =$ n-th coefficient group of the theory h. The preceding isomorphism shows that each element of $E_0^{p,q}$ may be identified with a p-cochain of X with values in $\check{h}^{q-p}$

(3. 8) $$E_0^{p,q} \cong C^p(X; \check{h}^{p-q}).$$

The differential d_0 is defined by the following diagram,

(3. 9) $$\begin{array}{ccccc} h^q(X_p, X_{p-1}) & \xrightarrow{\ \gamma\ } & h^q(X_p) & \xrightarrow{\ \beta\ } & h^{q+1}(X_{p+1}, X_p) \\ \cong \big\downarrow & & & & \big\downarrow \cong \\ C^p(X; \check{h}^{q-p}) & & \xrightarrow{\qquad d_0 \qquad} & & C^{p+1}(X; \check{h}^{q-p}) \end{array}$$

(3. 10) **Theorem.** Under the isomorphism $E_0^{p,q} \cong C^p(X;\check{h}^{q-p})$ the differential d_0 is identified to the ordinary cellular coboundary.

(3. 11) **Corollary.** $E_1^{p,q} \cong H^p(X; \check{h}^{q-p})$.

Proof of the theorem. The proof goes by a series of reductions. In fact, X/X_{p-1} is a complex whose (p-1)-skeleton is reduced to a point; it follows by naturality that we may replace X by X/X_{p-1} in studying d_0 on $C^p(X;\check{h}^{q-p})$, so we may assume that X_{p-1} is a point.

So, taking $X_{p-1} = \{pt\}$, one is reduced to looking at the case $X_p = \vee_\alpha S^p_\alpha$, $X_{p+1} = X_p \cup \{e^{p+1}_\beta\}$. Let $f : \vee_\beta S^p_\beta \to \vee_\alpha S^p_\alpha$ be the attaching map of the (p+1)-cells in X_{p+1} and consider the mapping cone sequence:

$$\begin{array}{ccccccccc}
\vee_\beta S^p_\beta & \xrightarrow{f} & \vee_\alpha S^p_\alpha & \xrightarrow{i} & C_f & \xrightarrow{j} & C_i & \xrightarrow{\Sigma f} & C_j \\
 & & \| & & \| & & \wr\| & & \wr\| \\
 & & X_p & \xrightarrow{i} & X_{p+1} & \longrightarrow & X_{p+1}/X_p & \longrightarrow & \Sigma X_p \\
 & & & & & & \| & \nearrow{\scriptstyle \Sigma f} & \\
 & & & & & & \Sigma(\vee_\beta S^p_\beta) & &
\end{array}$$

This shows that d_0 is given by de-suspending $h^{q+1}(X_{p+1}/X_p) \xleftarrow{(\Sigma f)^*} h^{q+1}(\Sigma X_p)$.

Thus $d_0 = f^* : h^q(X_p) \to h^q(Y)$, where $Y = \vee_\beta S^p_\beta$.

Thus, given $f : Y \to X_p$ we must establish the commutativity of the diagram

$$(3.12) \qquad \begin{array}{ccc}
h^q(X_p) & \xrightarrow{f^*} & h^q(Y) \\
\cong \downarrow & & \downarrow \cong \\
C^p(X_p;\check{h}^{q-p}) & \xrightarrow{f^\#} & C^p(Y;\check{h}^{q-p})
\end{array} .$$

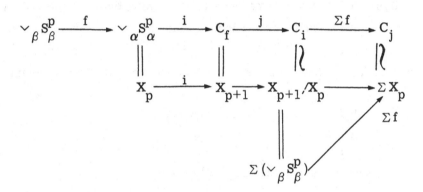

Second reduction. It suffices to look at the case $Y = S^p$. For, since, by (3. 7), $h^q(\vee_\beta S^p_\beta) \cong \Pi_\beta h^q(S^p_\beta)$, we may effect the determination of f^* and $f^\#$ by looking at them component by component.

Third reduction. In view of the previous reduction $Y = S^p$, compactness permits factoring $f:Y \to X_p$ into $Y \to X'_p \subseteq X_p$, where $X'_p = S^p_1 \vee \ldots \vee S^p_k$ is a finite subcomplex of X_p. Thus we may suppose X_p to be a finite wedge of spheres. Again, the analysis of f^* and $f^\#$ reduces to examining them component by component, so that we may indeed take X_p to be a single sphere.

Take $X_p = S^p$. Let $f:S^p \to S^p$ have degree d, then $f^\#$ will be multiplication by d. In order to prove that the diagram

$$\begin{array}{ccc} h^q(S^p) & \xleftarrow{f^*} & h^q(S^p) \\ \cong\downarrow & & \cong\downarrow \\ C^p(S^p;\check{h}^{q-p}) & \xleftarrow{f^\#} & C^p(S^p;\check{h}^{q-p}) \end{array}$$

is commutative, it suffices to prove that $f^*:h^q(S^p) \to h^q(S^p)$ is also multiplication by d.

In the cases $d = 0, 1$ this is clear from the Hopf classification theorem for maps of spheres and the homotopy axiom for h. Suppose $d > 1$. A map of degree d may be obtained by composition,

$$f \simeq w \circ g, \quad S^p \xrightarrow{g} \underbrace{S^p \vee S^p \vee \ldots \vee S^p}_{\text{(d times)}} \xrightarrow{w} S^p$$

where g is of degree +1 on each sphere and w is the folding map. Then $f^* = g^* w^*$ may be analysed as follows. $w^*:h^q(S^p) \to$

$\prod_{i=1}^{d} h^q(S_i^p)$ is given by $u \longmapsto (u, u, \ldots, u)$, while h^* takes each of the elements $(u, 0, \ldots, 0), (0, u, \ldots, 0), \ldots, (0, 0, \ldots, u)$ into $u \in h^q(S^p)$. To verify this last assertion, consider the case $h^*(u, 0, \ldots, 0) = u$. It suffices to observe that if $r: S_1^p \vee \ldots \vee S_d^p \to S_1^p$ is the retraction on the first sphere, then $r^*(u) = (u, 0, \ldots, 0)$ and the homotopy commutativity of the diagram,

$$\begin{array}{ccc} S^p & \xrightarrow{h} & S_1^p \vee \ldots \vee S_d^p \\ & \searrow & \downarrow r \\ rh \simeq 1 & & S_1^p \end{array}$$

gives us $h^* r^*(u) = u$ or $h^*(u, 0, \ldots, 0) = u$. We argue similarly for the other components. Consequently,

$$\begin{aligned} f^*(u) &= h^*(w^*(u)) = h^*(u, u, \ldots, u) \\ &= h^*((u, 0, \ldots, 0) + \ldots + (0, 0, \ldots, u)) \\ &= u + u + \ldots + u = du . \end{aligned}$$

It remains, finally, to look at the case $d = -1$ since then all other cases of d negative follow. Look at the composite map,

$$S^p \xrightarrow{\ \phi\ } S^p \vee S^p \xrightarrow{\ w\ } S^p ,$$

where ϕ is of degree $+1$ on the first sphere and -1 on the second, w the folding map. Then, supposing f of degree -1, we have

$$\begin{aligned} \phi^* w^*(u) &= \phi^*((u, 0) + (0, u)) = \phi^*(u, 0) + \phi^*(0, u) \\ &= u + f^*(u) = 0 \end{aligned}$$

since $\phi \circ w \simeq 0$, thus completing the proof of (3.10).

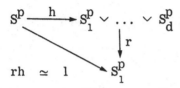

The remainder of this chapter will be devoted to the comparison theorem for spectral sequences and some of its applications.

If the space $X \in \mathscr{C}$ has dimension k, then the spectral sequence becomes stationary after a certain point. More precisely, we have the following lemma (the proof of which does not depend on the additional hypothesis (3. 7)).

(3. 13) **Lemma.** $E_k^{p,q} = E_{k+1}^{p,q} = \dots = E_\infty^{p,q}$, <u>where</u> $\dim X = k$.

Proof. Since $E_0^{p,q} = h^q(X_p, X_{p-1}) = h^q(X_p/X_{p-1})$, it follows that for $p < 0$ or $p > k$ we have $E_0^{p,q} = 0$, for in either case X_p/X_{p-1} reduces to a point. Thus $E_k^{p,q} = E_{k+1}^{p,q} = \dots = E_\infty^{p,q} = 0$ if $p < 0$ or $p > k$. It remains to verify the statement of the lemma for $0 \le p \le k$. Consider $E_n^{p-n-1,q-1} \xrightarrow{d_n} E_n^{p,q} \xrightarrow{d_n} E_n^{p+n+1,q+1}$; if $n \ge k$, then $p - n - 1 < 0$ and $p + n + 1 > k$ so that $E_{n+1}^{p,q} = \ker d_n / \operatorname{im} d_n = E_n^{p,q}$. Hence $E_k^{p,q} = E_{k+1}^{p,q} = \dots = E_\infty^{p,q}$.

In the proof of the Comparison Theorem, which follows, the additional hypothesis (3. 7) on the cohomology theory h is also not needed.

(3. 14) **Theorem.** <u>Consider two pairs</u> (X, h) <u>and</u> (X', h') <u>formed by spaces</u> $X, X' \in \mathscr{C}$ <u>and cohomology theories</u> h, h'. <u>Suppose that we have a map</u> ϕ <u>of the exact couple</u> $EC(X, h)$ <u>into the exact couple</u> $EC(X', h')$. <u>Then if for some</u> n, $\phi_n : E_n \cong E'_n$, <u>then</u> $\phi_* : h(X) \cong h'(X')$.

Proof. A map $\phi: EC(X,h) \to EC(X',h')$ is given by a family of morphisms (for each p)

$$\begin{array}{ccccccccc}
\cdots \to h^q(X_p, X_{p-1}) & \xrightarrow{\gamma} & h^q(X_p) & \xrightarrow{\alpha} & h^q(X_{p-1}) & \xrightarrow{\beta} & h^{q+1}(X_p, X_{p-1}) \to \cdots \\
\downarrow \phi & & \downarrow \phi & & \downarrow \phi & & \downarrow \phi \\
\cdots \to h'^q(X'_p, X'_{p-1}) & \xrightarrow{\gamma'} & h'^q(X'_p) & \xrightarrow{\alpha'} & h'^q(X'_{p-1}) & \xrightarrow{\beta'} & h'^{q+1}(X'_p, X'_{p-1}) \cdots
\end{array}$$

such that the whole diagram is commutative. The map ϕ of the exact couples induces a map of the spectral sequences. Each $\phi_n: E_n \to E'_n$ is a chain map (relative to d_n) and $\phi_{n+1} = (\phi_n)_*$, the map induced in homology. According to the preceding lemma the spectral sequences become stationary*, so that if, for some $n, \phi_n: E_n \to E'_n$ is an isomorphism, the same is true of $\phi_{n+1}, \phi_{n+2}, \ldots, \phi_\infty$. Let $m > p$; the diagram

$$\begin{array}{ccc}
h^q(X_m) & \longrightarrow & h^q(X_{p-1}) \\
\downarrow \phi & & \downarrow \phi \\
h'^q(X'_m) & \longrightarrow & h'^q(X'_{p-1})
\end{array},$$

obtained by iterating α and α', is commutative; and the fact that, for m sufficiently large, $X_m = X$ and $X'_m = X'$ shows that ϕ carries $F^p(h^q(X)) = \ker(h^q(X) \to h^q(X_{p-1}))$ into $F^p(h^q(X')) = \ker(h^q(X') \to h^q(X'_{p-1}))$. Let $k = \max(\dim X, \dim X')$; then we have a map of the filtrations,

---

* Actually the conclusion that ϕ_∞ is an isomorphism does <u>not</u> require stationarity.

$$\begin{array}{ccccccccc} 0 = F^{k+1}h^q(X) & \subseteq & F^k h^q(X) & \subseteq & \ldots & \subseteq & F^0 h^q(X) & = & h^q(X) \\ \downarrow \phi^{k+1} & & \downarrow \phi^k & & & & \downarrow \phi^0 & & \\ 0 = F^{k+1}h'^q(X') & \subseteq & F^k h'^q(X') & \subseteq & \ldots & \subseteq & F^0 h'^q(X') & = & h'^q(X') \end{array} .$$

In view of (3. 4) and (3. 5), we have the following commutative diagram

(3. 15) $$\begin{array}{ccccccccc} 0 & \longrightarrow & F^{p+1}h^q(X) & \longrightarrow & F^p h^q(X) & \longrightarrow & E_\infty^{p,q} & \longrightarrow & 0 \\ & & \downarrow \phi^{p+1} & & \downarrow \phi^p & & \downarrow \phi_\infty & & \\ 0 & \longrightarrow & F^{p+1}h'^q(X') & \longrightarrow & F^p h'^q(X') & \longrightarrow & E'^{p,q}_\infty & \longrightarrow & 0 \end{array}$$

Using (3. 15) and the fact that $\phi_\infty : E_\infty^{p,q} \cong E'^{p,q}_\infty$, we argue by induction that if ϕ^{p+1} is an isomorphism, the same is true for ϕ^p (using the five lemma). Certainly ϕ^{k+1} is an isomorphism because both groups are (0).

Hence $\phi^0 = \phi_* : F^0 h^q(X) = h^q(X) \cong F^0 h'^q(X') = h'^q(X')$ is an isomorphism, which proves the theorem.

As a consequence of the Comparison Theorem, we obtain the theorem below.

(3. 16) **Theorem.** If a map $f: X \to X'$ induces an isomorphism in ordinary homology, $f_* : H_*(X) \cong H_*(X')$, then f induces an isomorphism for any cohomology theory h, which satisfies the additional hypothesis (3. 7). (No additional hypothesis is necessary if X, X' are finite.)

Proof. We may assume that $f:X \to X'$ is cellular, i.e. $f(X_p) \subseteq X'_p$ for all p, since any map is homotopic to such a cellular map. The cellular map f induces a map of the exact couple $EC(X', h)$ into the exact couple $EC(X, h)$. In view of the commutativity of the diagram

$$\begin{array}{ccc} h^q(X'_p, X'_{p-1}) & \xrightarrow{f^*} & h^q(X_p, X_{p-1}) \\ \Vert\wr & & \Vert\wr \\ C^p(X'; \check{h}^{q-p}) & \xrightarrow{f^\#} & C^p(X; \check{h}^{q-p}) \end{array} ,$$

(this follows from (3.12)) and Corollary (3.11), the induced map at the E_1 level is just

$$f^*: H^p(X'; \check{h}^{q-p}) \to H^p(X; \check{h}^{q-p}) .$$

Using universal coefficients and the hypothesis on homology, the commutative diagram

$$\begin{array}{ccccccccc} 0 & \leftarrow & \mathrm{Hom}(H_p(X'), \check{h}^{q-p}) & \leftarrow & H^p(X'; \check{h}^{q-p}) & \leftarrow & \mathrm{Ext}(H_{p-1}(X'), \check{h}^{q-p}) & \leftarrow & 0 \\ \downarrow & & \downarrow \cong & & \downarrow f^* & & \downarrow \cong & & \downarrow \\ 0 & \leftarrow & \mathrm{Hom}(H_p(X), \check{h}^{q-p}) & \leftarrow & H^p(X; \check{h}^{q-p}) & \leftarrow & \mathrm{Ext}(H_{p-1}(X), \check{h}^{q-p}) & \leftarrow & 0 \end{array}$$

and the five-lemma implies $f^*: H^p(X'; \check{h}^{q-p}) \cong H^p(X; \check{h}^{q-p})$ for all p and q. Hence the map of exact couples defined by f induces an isomorphism $E'_1 \xrightarrow{\cong} E_1$; the proof is completed by applying the Comparison Theorem (3.14).

Remark. If X and X' are simply connected, the hypothesis that $f_*: H_*(X) \cong H_*(X')$ implies, by Whitehead's theorem, that f is a homotopy equivalence and the conclusion of (3.16), $f^*:h(X') \cong h(X)$ follows readily from the axioms. In the non-simply connected case, the preceding theorem may be interpreted loosely as asserting that as far as the additive structure of cohomology is concerned, general cohomology will not distinguish between spaces which ordinary cohomology will not distinguish. Actually we may use Whitehead's theorem even in the non-simply connected case, since the double suspension of any complex is simply connected. We then get a stronger result than (3.16).

For a second application of the comparison theorem, we need a definition. By a natural transformation of cohomology theories $t:h \to h'$ we mean a family of homomorphisms $t_X:h(X) \to h'(X)$ such that for every map $f:X \to Y$ the diagram

$$\begin{array}{ccc} h(X) & \xrightarrow{t_X} & h'(X) \\ {\scriptstyle h(f)}\uparrow & & \uparrow \\ h(Y) & \xrightarrow{t_Y} & h'(Y) \end{array}$$

is commutative; also, functoriality with respect to suspension σ is required, i.e. commutativity holds in the diagram

$$\begin{array}{ccc} h^n(X) & \xrightarrow{t_X} & h'^n(X) \\ {\scriptstyle \sigma}\downarrow{\scriptstyle \cong} & & {\scriptstyle \sigma'}\downarrow{\scriptstyle \cong} \\ h^{n+1}(\Sigma X) & \xrightarrow{t_{\Sigma X}} & h'^{n+1}(\Sigma X) \end{array} .$$

For $X = S^0$, we denote t_X by $\breve{t}:\breve{h} \to \breve{h}'$.

(3.17) **Theorem.** _If_ h _and_ h' _are cohomology theories defined on the category_ $\mathscr{C}$ _and_ $t:h \to h'$ _is a natural transformation of cohomology theories such that_ $\check{t}:\check{h} \cong \check{h}'$, _then_ $t:h \cong h'$.

Proof. In view of the definition, t certainly induces a map of the corresponding exact couples, $t:EC(X,h) \to EC(X',h')$. At the E_0 level, we get

$$\begin{array}{ccc} E_0^{p,q} & \xrightarrow{\ t\ } & E'^{p,q}_0 \\ \|\wr & \check{t}_* & \|\wr \\ C^p(X;\check{h}^{q-p}) & \longrightarrow & C^p(X;\check{h}'^{q-p}) \end{array}$$

where the bottom map is the coefficient homomorphism for cochains induced by the isomorphism $\check{t}:\check{h} \cong \check{h}'$; therefore $\check{t}_*$ is an isomorphism. Hence $t:E_0^{p,q} \cong E'^{p,q}_0$ for all p, q, or $t:E_0 \cong E'_0$. Now apply the comparison theorem to conclude that $t:h(X) \cong h'(X)$.

It is very important to notice that Theorem (3.17) should _not_ be interpreted as saying that a cohomology theory is determined by its coefficients. One certainly has distinct cohomology theories with the same coefficients. It is essential for the conclusion of the theorem that the isomorphism of coefficients is induced by a natural transformation of theories.

Theorem (3.17) has, in turn, an application which gives an interesting representation theorem in terms of stable cohomotopy, provided a suitable restriction is made on the coefficient group $\check{h}$. Let Q = field of rational numbers.

(3.18) **Theorem.** Let h be a cohomology theory such that $\check{h}$ is a graded vector space over Q, then

$$h^q(X) \cong \bigoplus_{r+s=q} \pi^r(X) \otimes \check{h}^s ,$$

where $\pi^r(X)$ stands for the rth (stable) cohomotopy group of X.

Proof. Let $h'^q(X) = \bigoplus_{r+s=q} \pi^r(X) \otimes \check{h}^s$; then it is not difficult to verify that h' is a cohomology theory. If $f:X \to Y$ is a mapping then $h'^q(f)$ is defined as a direct sum of homomorphisms $\pi^r(f) \otimes \mathrm{id}: \pi^r(Y) \otimes \check{h}^s \to \pi^r(X) \otimes \check{h}^s$. Invariance under homotopy and isomorphism under suspension follow easily. The exactness property follows by tensoring the exactness of π^r with $\check{h}^s$; $\check{h}^s$ being a locally free abelian group, exactness is preserved.

Now we set up a natural transformation $t:h' \to h$ between these cohomology theories. An element $\alpha \in \pi^r(X) = \lim_{\vec{n}}[\Sigma^n X, S^{n+r}]$ is represented by a map $f:\Sigma^n X \to S^{r+n}$; another map $g:\Sigma^m X \to S^{r+m}$ represents the same element if, for some $k \geq 0$,

$$\Sigma^{m+k} f, \quad \Sigma^{n+k} g: \Sigma^{m+n+k} X \to S^{r+m+n+k} \quad \text{are homotopic maps.}$$

Let $\beta \in \check{h}^s$, where $q = r + s$; using suspension, $\beta \in \check{h}^s = h^s(S^0) \cong h^{n+r+s}(S^{n+r})$. Now define $t(\alpha \otimes \beta) = f^*(\beta) \in h^{n+q}(\Sigma^n X) \cong h^q(X)$. That the definition does not depend on the choice of representative f follows from the fact that if g is another representative, then, for some $k \geq 0$, $\Sigma^{m+k} f \simeq \Sigma^{n+k} g$; passing to cohomology and de-suspending we get the desired conclusion. Bilinearity also follows by looking at track addition in $[\Sigma^{m+n+k} X, S^{r+m+n+k}]$.

To check the naturality of t thus defined is now a routine procedure. We will show that t is a natural equivalence.

The proof will consist of an application of (3.17); for this we must examine the natural transformation on the coefficients, i.e. $\check{t}:\check{h}' \to \check{h}$. Recall that

$$\pi^r(S^0) = \text{stable } (-r)\text{-stem} = \begin{cases} 0 \text{ if } r > 0, \\ Z \text{ if } r = 0 \\ \text{finite group if } r < 0. \end{cases}$$

Since the rationals Q form a divisible group, the tensor product (finite group) $\otimes$ Q = 0. So $h'^q = \bigoplus_{r+s=q} \pi^r(S^0) \otimes \check{h}^s =$

$\pi^0(S^0) \otimes \check{h}^q = Z \otimes \check{h}^q = \check{h}^q$. Moreover, $\check{t}$ is the identity. For let $\alpha \epsilon \pi^0(S^0) = Z$ be a generator; a representative of α is given by $f = id:S^{n+k} \to S^{n+k}$ and $f^*(\beta) = \beta$ so that $t(\alpha \otimes \beta) = t(1 \otimes \beta) = \beta$. That is, with the foregoing identification, $t:h'^q \to h^q$ is the identity isomorphism. Now we apply (3.17) to conclude that $t:h' \cong h$, whence

(3.19) $$h^q(X) \cong \bigoplus_{r+s=q} \pi^r(X) \otimes \check{h}^s .$$

Remarks. 1) In particular, ordinary reduced cohomology with rational coefficients $H^q(X;Q)$ satisfies the hypothesis of (3.18); the coefficient group $\check{H} = \check{H}^0 = Q$. In this case (3.19) becomes

$$H^q(X;Q) \cong \pi^q(X) \otimes Q .$$

2) The isomorphism of (3.19) is natural, i.e. it is realized by a natural transformation.

On the other hand, consider the cohomology theory $h''^q(X) = \bigoplus_{r+s=q} H^r(X;Q) \otimes \check{h}^s$, where H^r denotes ordinary reduced cohomology. Since $H^r(S^0;Q) = Q$ if $r = 0$ and (0) if $r \neq 0$, it follows that $\check{h}''^q = Q \otimes \check{h}^q = \check{h}^q$ if $\check{h}^q$ is a vector space over Q. Consequently, if h satisfies the hypothesis of (3.18) then h'' also does and $h''^q(X) \cong \bigoplus_{r+s=q} \pi^r(X) \otimes \check{h}^s \cong h^q(X)$. We conclude that if $\check{h}$ is a graded vector space over Q, then

$$(3.20) \qquad h^q(X) \cong \bigoplus_{r+s=q} H^r(X;Q) \otimes \check{h}^s .$$

Now let h be an arbitrary cohomology theory which satisfies the additional hypothesis (3.7). Then $h \otimes Q$ is also a cohomology theory with coefficients $\check{h} \otimes Q$, which is a graded vector space over Q. Thus we have

$$(3.21) \qquad h^q(X) \otimes Q \cong \bigoplus_{r+s=q} H^r(X;Q) \otimes Q \cong \bigoplus_{r+s=q} H^r(X;Q) \otimes \check{h}^s .$$

Remark. The foregoing argument shows that the essential, unique features of a cohomology theory h are associated with torsion in $h(X)$.

We wish now to show that the isomorphisms (3.19), (3.20), (3.21) are uniquely determined by naturality and the requirement that they be the identity on the coefficients. In fact, we will prove a more general theorem.

(3.22) **Theorem.** (i) _If_ $\check{h}$ _is a graded vector space over_ Q, _then any natural transformation_ $T: h \to h'$ _is determined by_ $\check{T}: \check{h} \to \check{h}'$.

(ii) _If_ $\check{h}'$ _is a graded vector space over_ Q, _then any morphism_ $\check{T}: \check{h} \to \check{h}'$ _has an unique extension to a natural transformation_ $T: h \to h'$.

Proof. (i) According to Thm. (3.18) there exists an isomorphism $\bigoplus_{r+s=q} \pi^r(X) \otimes \check{h}^s \xrightarrow{\cong} h^q(X)$, given by $(\alpha \otimes \beta) \mapsto \alpha^*(\beta)$, writing α^* since the dependence is on the class α. Thus every element of $h^q(X)$ may be expressed as a sum of such elements $\alpha^*(\beta)$, so it suffices to show that T is determined on such elements. Now this follows easily using naturality and the definition of $\check{T}$, for

$$T(\alpha^*(\beta)) = \alpha^*(T\beta) = \alpha^*(\check{T}\beta) .$$

(ii) By the hypothesis on $\check{h}'$ and (3.18), we may assume that $h'^q(X) = \bigoplus_{r+s=q} \pi^r(X) \otimes \check{h}'^s$. Define a natural transformation T extending $\check{T}$ by $T: h^q(X) \to h'^q(X)$ using the following diagram:

$$\begin{array}{ccc} h^q(X) \longrightarrow h^q(X) \otimes Q \cong & & \bigoplus_{r+s=q} \pi^r(X) \otimes \check{h}^s \otimes Q \\ T \downarrow & & \downarrow \check{T} \\ h'^q(X) = \bigoplus_{r+s=q} \pi^r(X) \otimes \check{h}'^s & \xleftarrow{\cong} & \bigoplus_{r+s=q} \pi^r(X) \otimes (\check{h}'^s \otimes Q) , \end{array}$$

where $h^q(X) \to h^q(X) \otimes Q$ is $u \longmapsto u \otimes 1$, $\check{T}$ is $(id) \otimes (\check{T}^s \otimes (id))$ on each summand and the isomorphism in the lower row is induced by the natural isomorphism $\check{h}'^{s} \otimes Q \cong \check{h}'^{s}$. There remains to check the uniqueness of T. Any natural transformation $T: h \to h'$ may be factored as follows,

$$\begin{array}{ccc} h^q(X) & \longrightarrow & h^q(X) \otimes Q \\ & {\scriptstyle T} \searrow \quad \swarrow {\scriptstyle T'} & \\ & h'^q(X) & \end{array}$$

where $T'(u \otimes \rho) = T(u).\rho \in h'^q(X)$, $\rho \in Q$; every T gives rise to a natural transformation T' and, in turn, is uniquely determined by T'. Now suppose that $T_1, T_2 : h \to h'$ are natural transformations such that $\check{T}_1 = \check{T}_2$, then in the diagram

$$\begin{array}{ccc} \check{h}^q & \longrightarrow & \check{h}^q \otimes Q \\ & {\scriptstyle \check{T}_1 = \check{T}_2} \searrow \quad \swarrow {\scriptstyle \check{T}'_1,\ \check{T}'_2} & \\ & \check{h}'^q & \end{array}$$

$T'_1(u \otimes \rho) = \check{T}_1(u).\rho = \check{T}_2(u).\rho = \check{T}'_2(u \otimes \rho)$. Applying part (i) to $T'_1, T'_2 : h^q(X) \otimes Q \to h'^q(X)$, we conclude that $T'_1 = T'_2$, whence $T_1 = T_2$.

We refer to the natural transformation $h \to h \otimes Q$ as the <u>character</u> of the theory h.

To close this chapter, we prove the uniqueness theorem for ordinary cohomology and a theorem of finitude for general cohomology.

(3.23) **Theorem.** <u>Let</u> h <u>be any cohomology theory satisfying</u> (3.7) <u>and the dimension axiom, i.e.</u>

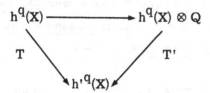

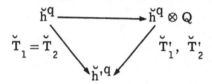

$\check{h}^0 = G$ and $\check{h}^q = 0$ for $q \neq 0$, then $h^q(X) \cong H^q(X;G)$.

Proof. According to Corollary (3.12) the E_1 term of the spectral sequence is given by

$$E_1^{p,q} = H^p(X; \check{h}^{q-p}) = \begin{cases} H^q(X;G) & \text{if } p = q \\ 0 & \text{if } p \neq q . \end{cases}$$

Recall that the bidegree of d_n is $(n+1, 1)$; it follows that for $n \geq 1$, $d_n = 0$ and hence that $E_r^{p,q}$ becomes stationary for $r \geq 1$. Hence,

$$E_\infty^{p,q} \cong \begin{cases} H^q(X;G) & \text{if } p = q; \\ 0 & \text{if } p \neq q . \end{cases}$$

Let $k = \dim X$ and look at the filtration

$$0 = F^{k+1}h^q(X) \subseteq F^k h^q(X) \subseteq \ldots \subseteq F^0 h^q(X) = h^q(X) .$$

Using the above value for $E_\infty^{p,q}$ and (3.6), namely,

$$F^p h^q(X) / F^{p+1} h^q(X) \cong E_\infty^{p,q} ,$$

we see that the filtration reduces to,

$$0 \subseteq \ldots \subseteq 0 \subseteq F^q h^q(X) = H^q(X, G) = \ldots = F^0 h^q(X) = h^q(X),$$

that is, $h^q(X) = H^q(X;G)$.

As an application, we deduce an important corollary. Let $\underline{Y} = \{Y_k\}$ be the Eilenberg-MacLane spectrum, i.e. $Y_n = K(G, n)$ for $n \geq 0$, $Y_n = pt$ for $n < 0$. Consider the cohomology theory h_Y (see chapter I) defined by $h_Y^q(X) = [X, Y_q]$. Now h_Y satisfies the dimension axiom: for $q \neq 0$, we have

$$\check{h}_Y^q = h_Y^q(S^0) = [S^0, Y_q] = 0,$$

since, for $q > 0$, Y_q is pathwise connected. For $q = 0$,

$$\check{h}_Y^q = h_Y^0(S^0) = [S^0, G] \cong G .$$

Applying Theorem (3.23), we conclude the classical result of Serre:

(3.24) **Corollary.** $h_Y^q(X) = [X, Y_q] \cong H^q(X, G)$.

Remark. Theorem (3.23) establishes the uniqueness theorem for cohomology theories defined on the category of finite complexes and satisfying the dimension axiom; it also establishes the uniqueness theorem for such theories defined on the category of finite-dimensional complexes, provided the theory also satisfies (3.7). If the theory is not required to satisfy (3.7) it is easy to give a counterexample to uniqueness.

(3.25) **Theorem.** Let X be a finite cell complex, h a cohomology theory satisfying (3.7) and such that $\check{h}^q$ is finitely generated for each q. Then $h^q(X)$ is finitely generated for each q.

Proof. According to (3. 8), $E_0^{p,q} \cong C^p(X; \check{h}^{q-p})$, which is finitely generated according to the hypothesis. Hence $E_\infty^{p,q}$ is finitely generated. Using Theorem (3. 6) and arguing by induction on the filtration degree, we conclude that $F^0h^q(X) = h^q(X)$ is finitely generated.

The reader will notice that this argument generalizes to other properties of the coefficient groups $\check{h}^q$ and shows such properties to hold for $h^q(X)$, **X** finite, if they hold for the coefficients.

4. K-Theory

We have already given a brief description of (complex) K-theory in terms of the unitary spectrum. We now give an alternative, geometrical description which serves better to explain its role in topology. Let E and X be given spaces and let p be a map of E onto X. We are going to make several assumptions about the triple (E, p, X). To begin with, we shall assume that, for every $x \in X$, $p^{-1}(x)$ has the structure of a finite-dimensional complex vector space. If the vector space associated to the fibre $p^{-1}(x)$ has dimension n, we shall indicate it by C^n, where C denotes the field of complex numbers.

Secondly, we suppose that X is covered by a family of open subsets U_α and that, to each U_α, we are given a homeomorphism

$$\psi_\alpha : U_\alpha \times C^n \longrightarrow p^{-1}(U_\alpha)$$

such that $p\psi_\alpha(x, v) = x$, for every $(x, v) \in U_\alpha \times C^n$, and ψ_α is consistent with the vector space structure in the fibres. Observe that if $x \in U_\alpha \cap U_\beta$, $\psi_\beta^{-1} \psi_\alpha(x, v) = (x, v')$. Thus, given x, to a vector $v \in C^n$ we associate the vector v', also in C^n; let us write $v' = [\rho_{\alpha\beta}(x)]$ (v). It then follows that this correspondence between v and v' is actually given by a continuous function $\rho_{\alpha\beta} : U_\alpha \cap U_\beta \to GL(C, n)$, the general linear group.

A triple (E, p, X) satisfying the conditions listed above is called a complex vector-bundle over X. (There is, evidently, an analogous notion of real vector-bundle.)

Observe that the dimension of the vector space $p^{-1}(X)$ is locally constant on X and therefore, is constant on the connected components of X. If, for every $x \in X$, the dimension of $p^{-1}(X)$ is equal to a fixed positive integer n, we say that (E, p, X) is a complex n-bundle over X.

As an example of a complex n-bundle, let us take the triple $(X \times C^n, p, X)$ where p is the projection on the first factor. This is the trivial n-bundle over X. A second example is the tangent bundle of a complex n-manifold.

Notice that a general complex vector bundle over X is locally trivial.

Given a complex vector bundle (E, p, X) and a continuous function $f:Y \to X$ we define a complex vector bundle over Y as follows: let $F = \{(y, e) \in Y \times E \mid f(y) = p(e)\}$ and let $\pi_1;F \to Y$ be the projection on the first factor. The space Y is covered by the open sets $f^{-1}(U_\alpha)$ and the continuous function $\bar{\rho}_{\alpha\beta}:f^{-1}(U_\alpha) \cap f^{-1}(U_\beta) \to GL(C, n)$ is defined by $[\bar{\rho}_{\alpha\beta}(y)](v) = [\rho_{\alpha\beta}(f(y))](v)$, for every $y \in f^{-1}(U_\alpha) \cap f^{-1}(U_\beta)$ and every $v \in C^n$. The triple (F, π_1, Y) is a vector bundle over Y, called the bundle induced by f from (E, p, X). If we denote the bundle (E, p, X) by ξ, the induced bundle (F, π_1, Y) will be denoted by $f^*(\xi)$.

Two bundles over X, say (E_1, p_1, X) and (E_2, p_2, X) are equivalent if, and only if, there exists a homeomorphism $\phi:E_1 \to E_2$ such that $p_2\phi = p_1$ and, when restricted to a fibre $p_1^{-1}(x)$, ϕ is an isomorphism of $p_1^{-1}(x)$ onto $p_2^{-1}(x)$. This is an equivalence relation in the set of all vector bundles over X. We shall denote by B(X) the set of all equivalence classes of complex vector bundles over X. Let $B_k(X)$ be the subset of equivalence classes of k-bundles.

One should observe that if $f:Y \to X$ is a given map, and if ξ and ξ' are equivalent vector bundles over X, the induced bundles $f^*(\xi)$ and $f^*(\xi')$ are equivalent vector bundles over Y. Hence a map $f:Y \to X$ induces a function f^* of the set B(X) into the set B(Y), sending $B_k(X)$ to $B_k(Y)$.

We shall introduce an addition operation between vector bundles over the same space, so that the sets B(X) and B(Y) become semi-groups with zero element, and f^* becomes a semi-group homomorphism.

Given the vector bundles $\xi_1 = (E_1, p_1, X)$ and $\xi_2 = (E_2, p_2, X)$ we define in an evident way the bundle $\xi_1 \times \xi_2 = (E_1 \times E_2, p_1 \times p_2, X \times X)$; the diagonal map $d:X \to X \times X$ which takes any $x \in X$ into (x, x) then induces a bundle $\xi_1 \oplus \xi_2 = d^*(\xi_1 \times \xi_2)$ over X. The bundle $\xi_1 \oplus \xi_2$ is called the Whitney sum bundle of ξ_1 and ξ_2. We shall give next another definition of the Whitney sum in order to get a better geometric insight into it.

The bundles ξ_1 and ξ_2 are locally trivial, that is to say, the space X is covered by two families of open sets $\{U^1_\alpha\}$ and $\{U^2_\beta\}$ such that the part of ξ_1 which lies over an open set U^1_α is trivial and the part of ξ_2 lying over an open set U^2_β is a trivial bundle. By taking intersections between the two families, we can obviously cover X by a family $\{V_\gamma\}$ of open sets such that both ξ_1 and ξ_2 have the local triviality property with respect to this open cover. Define the set $E_1 \oplus E_2 =$ $\{(x, u \oplus v) \in X \times (p_1^{-1}(x) \oplus p_2^{-1}(x)), x \in X\}$ and the map $p:E_1 \oplus E_2 \to X$ to be the projection on the first factor. If $\psi^1_\gamma : V_\gamma \times C^n \to p_1^{-1}(V_\gamma)$ and $\psi^2_\gamma : V_\gamma \times C^m \to p_2^{-1}(V_\gamma)$ are the local homeomorphisms, define $\psi_\gamma : V_\gamma \times (C^n \oplus C^m) \to p^{-1}(V_\gamma)$

by $\psi_\gamma(x, (u \oplus v)) = (x, \psi^1_\gamma(x, u) \oplus \psi^2_\gamma(x, v))$. The bundle obtained by suitably topologizing $E_1 \oplus E_2$ is equivalent to the Whitney sum introduced before. Speaking less accurately, the bundle $\xi_1 \oplus \xi_2$ is obtained by replacing the fibres $p_1^{-1}(x)$ and $p_2^{-1}(x)$ over x by their direct sum.

If in the previous process, instead of using direct sums we take tensor products, the complex vector bundle attached to ξ_1 and ξ_2 would be the tensor product bundle $\xi_1 \otimes \xi_2$.

Notice that if ϵ_0 is the complex 0-bundle over X, then for any complex vector bundle ξ over X, the Whitney sums $\xi \oplus \epsilon_0$ and $\epsilon_0 \oplus \xi$ are equivalent to the bundle ξ. If ϵ_1 is the trivial 1-bundle over X, the bundles ξ, $\xi \otimes \epsilon_1$ and $\epsilon_1 \otimes \xi$ are equivalent. Also, the tensor product of bundles is distributive over the Whitney sum.

Finally, given the vector bundle ξ over X, since its fibre over any $x \in X$ is a complex vector space, we can obtain a new bundle $\Lambda^r \xi$ by replacing the fibres of ξ by their exterior r^{th}-powers. It is important to observe that the above operations on bundles preserve vector bundle equivalence, thus they pass to B(X). (For more detailed study of operations on vector bundles, the reader is referred to: Husemoller, D., Fibre Bundles, McGraw-Hill, Chapter 5, Section 6.)

Let $G_{k,n}$ be the complex Grassmann Manifold of k-planes through the origin in the complex vector space C^n, that is to say, the set of all k-dimensional vector subspaces of C^n, $k \leq n$. We define the space $E_{k,n}$ to be the set of all pairs $(V, v) \in G_{k,n} \times C^n$ such that $v \in V$, and set $\pi : E_{k,n} \to G_{k,n}$ to be the projection on the first factor. The triple $\gamma_{k,n} = (E_{k,n}, \pi, G_{k,n})$ is a complex k-bundle. This canonical k-bundle $\gamma_{k,n}$ has the following important property:

(4.1) **Classification Theorem.** Let X be a finite dimensional cell complex of dimension r. The function that assigns to the homotopy class of a map $f:X \to G_{k,n}$ the complex k-bundle $f^*(\gamma_{k,n})$ over X is a bijection of $[X, G_{k,n}]$ with $B_k(X)$, whenever $r \leq 2(n - k)$ (Husemoller, D., Fibre Bundles (McGraw-Hill) Chapter 7, Theorem 7.2).

We now turn to the construction of K(X); we first need the notion of a Grothendieck group. We shall show how to every abelian semi-group with zero element S it is possible to associate an abelian group G(S) - the Grothendieck Group of S - and a semi-group homomorphism $\gamma:S \to G(S)$ satisfying the following universal property:

(4.2) if A is an arbitrary abelian group and ϕ is a semi-group homomorphism from S into A, there exists a unique group homomorphism $\bar{\phi}:G(S) \to A$ which renders commutative the diagram

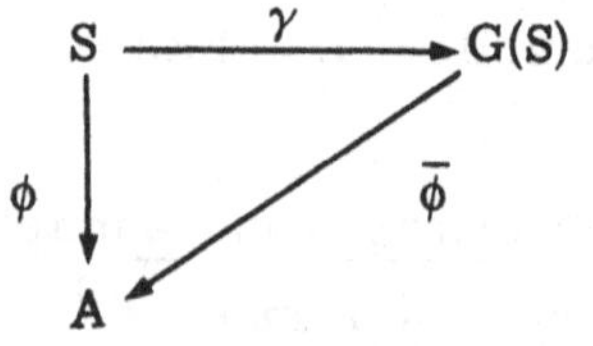

Before proving that such a group G(S) does exist, observe that, assuming its existence, G(S) must be unique up to canonical group isomorphism.

The group $G(S)$ can be obtained in the following way; define the pairs (a, b), $(c, d) \in S \times S$ to be equivalent if, and only if, there is $u \in S$ such that $a + d + u = b + c + u$. This is an equivalence relation in $S \times S$; let $G(S)$ be the quotient of $S \times S$ by this relation. We denote by $\{a, b\}$ the class of the pair $(a, b) \in S \times S$; then we define the sum $\{a, b\} + \{c, d\} = \{a+c, b+d\}$. If 0 is the zero element of S, the element $\{0, 0\} \in G(S)$ acts as zero element; moreover any element $\{a, b\} \in G(S)$ has an inverse $\{b, a\}$. (Observe that $(0, 0)$ is equivalent to (a, a), for any $a \in S$; thus even if S does not have a zero element, we can still associate an abelian group $G(S)$ with it in this way.)

The semi-group homomorphism $\gamma: S \to G(S)$ is defined by the rule $\gamma(a) = \{a, 0\}$, for every $a \in S$. Finally, given $\phi: S \to A$ a semi-group homomorphism, where A is an abelian group, we define $\bar{\phi}\{a, b\} = \phi(a) - \phi(b)$. It is easy to check that $\bar{\phi}$ is a well-defined group homomorphism and is the only one such that $\bar{\phi}\gamma = \phi$.

If $h: S_1 \to S_2$ is a semi-group homomorphism, define the group homomorphism $\bar{h}^*: G(S_1) \to G(S_2)$ by writing $\bar{h}^*\{a, b\} = \{h(a), h(b)\}$, for every $\{a, b\} \in G(S_1)$. Notice that if $\gamma_1: S_1 \to G(S_1)$ and $\gamma_2: S_2 \to G(S_2)$ are the homomorphisms satisfying (4.2) then $\gamma_2 \circ h = \bar{h}^* \circ \gamma_1$.

We apply these ideas to the semi-group defined by the Whitney sum on the set $B(X)$ of all equivalence classes of complex vector bundles over X.

Given any space $X \in \mathscr{C}$ we associate with it an abelian group $K(X)$ defined as the Grothendieck group $G(B(X))$. If $f: Y \to X$ is a morphism of $\mathscr{C}$, the operation of inducing a vector bundle over Y defines a semi-group homomorphism

$f^*: B(X) \to B(Y)$ and, hence, a group homomorphism, which we also write $f^*: K(X) \to K(Y)$. It follows from the classification theorem (4.1) that $K(X)$ is an invariant of homotopy type; furthermore, K is a contravariant functor from the category $\mathscr{C}$ to the category of abelian groups and homomorphisms.

Let x_0 be the base-point of X and let $i: \{x_0\} \to X$ be the inclusion map and $c: X \to \{x_0\}$ be the constant map. Since $c \circ i$ takes the point x_0 into itself, $i^* \circ c^*$ is the identity homomorphism of $K(X)$ and thus, i^* is an epimorphism. Defining $\tilde{K}(X)$ as the kernel of i^*, we obtain a split exact sequence

$$0 \longrightarrow \tilde{K}(X) \longrightarrow K(X) \underset{c^*}{\overset{i^*}{\rightleftarrows}} K(\{x_0\}) \longrightarrow 0$$

and therefore, $K(X) \cong \tilde{K}(X) \oplus K(\{x_0\})$.

On the other hand, every vector bundle over the point x_0 is trivial; if to an n-bundle over x_0 we associate the positive integer n, we obtain a one-to-one correspondence between $B(\{x_0\})$ and the semi-group Z_+ of all positive integers. This shows that $K(\{x_0\}) \cong Z$, and therefore

(4.3) $\quad K(X) \cong \tilde{K}(X) \oplus Z$.

Next we give a direct characterization of $\tilde{K}(X)$ in terms of vector bundles over X.

Two vector bundles ξ and ξ' over X are said to be stably equivalent (notation; $\xi \sim \xi'$) if there are trivial bundles ϵ and ϵ' over X such that $\xi \oplus \epsilon$ and $\xi' \oplus \epsilon'$ are equivalent. Clearly, equivalent vector bundles are stably equivalent, so that stable equivalence may be regarded as an equivalence relation in $B(X)$. We denote by $[\xi]$ the stable class of ξ and by $T(X)$ the semi-group of all stable-equivalence classes of $B(X)$. The zero element of $T(X)$ is the stable class of a trivial bundle.

(4. 4) **Lemma.** The semi-group $T(X)$ is an abelian group.

Proof. We have to show that, given any bundle ξ, there exists η such that $\xi \oplus \eta$ is equivalent to a trivial bundle. Let us suppose first that $X \in \mathscr{C}$ is connected and $\dim X = r$. Because X is connected, all the fibres of a vector bundle over X have the same dimension.

Given a complex k-bundle ξ over X, we know from the Classification Theorem that for n large enough (that is to say, n satisfying the condition $r \leq 2(n - k)$) there exists a map $f: X \to G_{k,n}$ such that the vector bundle ξ is equivalent to the induced bundle $f^*(\gamma_{k,n})$, where $\gamma_{k,n}$ is the canonical k-bundle over $G_{k,n}$. Let $\eta_{k,n}$ be the vector $(n - k)$-bundle $(E'_{k,n}, \pi', G_{k,n})$ with $E'_{k,n} = \{(V, u) \in G_{k,n} \times C^n \,|\, u \in V^\perp\}$, where $V^\perp$ is the normal $(n - k)$-plane to V in C^n. Since $\gamma_{k,n} \oplus \eta_{k,n}$ is the trivial n-bundle over $G_{k,n}$ (Husemoller, D., Fibre Bundles, McGraw-Hill, Chapter 2, Example 4. 5) $f^*(\gamma_{k,n} \oplus \eta_{k,n}) = \xi \oplus f^*(\eta_{k,n})$ is equivalent to the trivial n-bundle over X.

If X is not connected, this procedure can be followed for each one of its connected components. This completes the proof.

We now observe that by associating to the equivalence class of any bundle ξ over X its stable class, we obtain a semi-group homomorphism $\phi: B(X) \to T(X)$. Then by (4. 2) there exists a unique homomorphism $\bar{\phi}: K(X) \to T(X)$ such that $\bar{\phi}\gamma = \phi$. Uniqueness of $\bar{\phi}$ shows that this homomorphism can be defined as follows: given $\{\xi, \eta\} \in K(X)$, let $\bar{\eta}$ be a vector bundle over X

such that $\eta \oplus \bar{\eta} = \epsilon$, a trivial vector bundle; then $\{\xi, \eta\} = \{\xi \oplus \bar{\eta}, \epsilon\}$. Set $\bar{\phi}\{\xi, \eta\} = [\xi \oplus \bar{\eta}]$.

(4.5) **Theorem.** The restriction $\bar{\phi} \mid \tilde{K}(X)$ is an isomorphism of $\tilde{K}(X)$ onto $T(X)$.

Proof. It is plainly sufficient to assume that $X \in \mathscr{C}$ is connected. Given any $[\xi] \in T(X)$, where $\dim \xi = n$ take $\{\xi, \epsilon_n\} \in K(X)$, with ϵ_n the trivial n-bundle over X. Since $i^*\{\xi, \epsilon_n\} = \{i^*\xi, i^*\epsilon_n\}$ and both $i^*\xi$ and $i^*\epsilon_n$ are trivial n-bundles over x_0, it follows that $\{\xi, \epsilon_n\} \in \ker i^*$ and hence, to $\tilde{K}(X)$. On the other hand $\bar{\phi}\{\xi, \epsilon_n\} = [\xi]$ and so $\bar{\phi} \mid \tilde{K}(X)$ is an epimorphism.

Suppose next that $\{\xi, \eta\} \in \tilde{K}(X)$ and $\bar{\phi}\{\xi, \eta\} = 0$. Then $\dim \xi = \dim \eta$ and there exists $\bar{\eta}$ such that $\eta \oplus \bar{\eta} = \epsilon$, $\bar{\phi}\{\xi, \eta\} = [\xi \oplus \bar{\eta}] = 0$. That is to say, there exist trivial bundles ϵ', ϵ'' such that $\xi \oplus \bar{\eta} \oplus \epsilon' = \epsilon''$. Hence $\{\xi, \eta\} = \{\xi \oplus \bar{\eta}, \epsilon X\} = \{\epsilon'', \epsilon' \oplus \epsilon\} = \{\epsilon'', \epsilon''\} = 0$.

Notice that the inverse isomorphism $T(X) \to \tilde{K}(X)$ is given by $[\xi] \longmapsto \{\xi, \epsilon_n\}$, where ξ is an n-bundle.

In view of the theorem, since $K(X) \cong \tilde{K}(X) \oplus Z$, we may regard $K(X)$ as the group of pairs $([\xi], m)$, where $[\xi]$ is the stable class of a vector bundle ξ over X and m is an integer. The isomorphism between $K(X)$ and this group of pairs is given by $\{\xi, \epsilon_k\} \longmapsto ([\xi], n-k)$, where ξ is an n-bundle.

We now proceed to establish the connection between $K(X)$ as defined in this chapter and the unitary spectrum. First, however, we should observe that the operation of tensor product of bundles gives to $K(X)$ the structure of a ring with unit element.

Now let $X \in \mathscr{C}$ be a given connected space. We have seen that a k-bundle over X is classified by a map of X into $G_{k,n}$, for n sufficiently large. On the other hand, the Grassmann manifolds $G_{k,n}$ can be viewed in a more algebraic fashion which is very useful. Let $U(n)$ be the group of all unitary transformations of C^n, that is to say, of all linear transformations of C^n which are length preserving. If C^k is a fixed k-subspace of C^n and C^{n-k} is its orthogonal complement, the subgroup of $U(n)$ which takes C^k on itself splits up into the direct product $U(k) \times U'(n-k)$, where $U(k)$ is the subgroup of $U(n)$ which leaves C^k pointwise fixed and $U'(n-k)$ is the subgroup which leaves C^{n-k} pointwise fixed. If one represents the elements of $U(n)$ by $n \times n$ unitary matrices the elements of $U(k)$ and $U'(n-k)$, regarded as subgroups of $U(n)$, can be represented respectively, by matrices of the form;

$$\underbrace{\begin{bmatrix} \sigma & 0 & \cdots & & 0 \\ 0 & 1 & & & \\ \vdots & & 1 & & \\ \vdots & & & \ddots & \\ 0 & & & & 1 \end{bmatrix}}_{n-k} \text{ and } \underbrace{\begin{bmatrix} 1 & & & & 0 \\ & 1 & & & \vdots \\ & & \ddots & & \vdots \\ & & & 1 & 0 \\ 0 & \cdots & & 0 & \sigma' \end{bmatrix}}_{k}, \quad \sigma \in U(k), \quad \sigma' \in U'(n-k).$$

The space $G_{k,n}$ is then identified to the quotient $U(n)/U(k) \times U'(n-k)$; moreover, γ_{kn} is then seen as the vector bundle associated with the principal bundle $W_{k,n} \to G_{k,n}$, where $W_{k,n}$ is the complex Stiefel manifold $U(n)/U'(n-k)$.

The map of $U(n)$ into $U(n+1)$ which takes any matrix $\sigma \in U(n)$ into the matrix $\begin{pmatrix} \sigma & 0 \\ 0 & 1 \end{pmatrix} \in U(n+1)$ induces an embedding of $G_{k,n}$ into $G_{k,n+1}$. Define $G_k = \varinjlim_n G_{k,n}$. Moreover, by taking any $\sigma \in U(n)$ into $\begin{pmatrix} 1 & 0 \\ 0 & \sigma \end{pmatrix} \in U(n+1)$, one obtains an imbedding of $G_{k,n}$ into $G_{k+1,n+1}$ and ultimately, (in view of the compatibility of the two embeddings described) of G_k into G_{k+1}. We define the space B_U as the direct limit $\varinjlim_k G_k = \varinjlim_k \varinjlim_n G_{k,n}$.

Obviously, for any finite-dimensional complex X, equivalence classes of k-bundles over X are classified by $[X, G_k]$.

Let us consider the set of all homotopy classes of maps of X into B_U. Let us first suppose X connected. Since X is finite dimensional, an element $a \in [X, B_U]$ is represented by a map of X into G_k, for some k; thus, to such an element we may attach a k-bundle over X. Furthermore, notice that if $f: X \to G_k$ and $f': X \to G_{k'}$ are two representatives of a, corresponding to the vector bundles ξ and ξ' over X, then for some $k'' \geq k, k'$, the maps f and f' will be homotopic when viewed as maps of X into $G_{k''}$. This means that $\xi \oplus \epsilon_{k''-k} = \xi' \oplus \epsilon_{k''-k'}$ that is to say, ξ and ξ' are stably equivalent. This shows that there exists a one-to-one correspondence between T(X) and $[X, B_U]$, and hence between $\tilde{K}(X)$ and $[X, B_U]$.

If we drop the assumption that X is connected, we have a one-to-one correspondence between $\tilde{K}(X)$ and $[X, B_U \times Z]$, since we may distinguish the components of X by $[X, Z]$. This argument shows that $B_U \times Z$ is a group (up to homotopy); then the correspondence above is a group isomorphism. Indeed, $B_U \times Z$ is homotopically equivalent to a loop space, namely ΩU, where $U = \lim U(n)$ (Bott, R.; The Space of Loops on a Lie Group, Michigan J. of Math. 5 (1958); 35-61).

For every $X \in \mathcal{C}$, we set $\tilde{K}^0(X) = \tilde{K}(X)$ and proceed to define $\tilde{K}$-groups for all integral dimensions. To start off, for every integer $n \geq 0$ we define

$$\tilde{K}^{-n}(X) = \tilde{K}(\Sigma^n X). \tag{4.6}$$

Because of a well-known "adjointness property" of the functors Σ and Ω, $\tilde{K}^{-n}(X) \cong [\Sigma^n X, B_U \times Z] \cong [\Sigma^n X, \Omega U] \cong [X, \Omega^{n+1} U]$. Now the homotopy equivalence $\Omega^2 U \simeq U$ (see Chapter 1) shows that for any $X \in \mathscr{C}$ and any integer $n \geq 0$, the groups $\tilde{K}^{-n}(X)$ and $\tilde{K}^{-n-2}(X)$ are isomorphic. We define $\tilde{K}^n(X)$ for every $n \in Z$, inductively by $\tilde{K}^n(X) \cong \tilde{K}^{n-2}(X)$.

In this way we obtain again the complex K-theory we have defined in Chapter 1, using the unitary spectrum.

For finite-dimensional cell-complexes X without base-point, define the non-reduced K-theory as for the general case developed in Chapter 1; set X^+ to be the disjoint union of X and a point, then write $K^n(X) = \tilde{K}^n(X^+)$. The coefficients of this theory are:

(4. 7) $$K^n(\{x_0\}) = \begin{cases} Z & \text{if } n \text{ is even} \\ 0 & \text{if } n \text{ is odd}. \end{cases}$$

Because of the periodicity $K^n(X) \cong K^{n-2}(X)$ it is convenient to introduce the Z_2-graded ring

(4. 8) $$K^*(X) = K^0(X) \oplus K^1(X).$$

Using (3. 25) we conclude

(4. 9) $K^*(X)$ is of finite type, provided X is a finite cell complex.

According to (3. 21) we have the following isomorphism

(4. 10) $$K^n(X) \otimes Q \cong \bigoplus_{r+s=n} H^r(X; Q) \otimes K^s(\{x_0\}),$$

and because of (4.7), we obtain

$$(4.11) \quad K^*(X) \otimes Q \cong \bigoplus_{n} H^n(X; Q) .$$

More specifically,

$$(4.12) \quad K^0(X) \otimes Q \cong \bigoplus_{n \text{ even}} H^n(X; Q) ,$$

$$K^1(X) \otimes Q \cong \bigoplus_{n \text{ odd}} H^n(X; Q) .$$

Thus the <u>character</u> of K-theory may be understood as a natural transformation χ,

$$(4.13) \quad \chi : K^0(X) \to \bigoplus_{n \text{ even}} H^n(X; Q) ,$$

$$\chi : K^1(X) \to \bigoplus_{n \text{ odd}} H^n(X; Q) .$$

5. K-Theory, the Chern Character and the Hopf Invariant Problem

Let $[\alpha] \in \pi_{2n-1}(S^n)$, $n \geq 2$, be represented by a function $\alpha: S^{2n-1} \to S^n$. Form the sequence of spaces

$$S^{2n-1} \xrightarrow{\ \alpha\ } S^n \xrightarrow{\ i\ } C_\alpha \xrightarrow{\ j\ } \Sigma S^{2n-1} \longrightarrow \cdots$$

and consider the long exact sequence induced by it in ordinary cohomology. The sequence of cohomology groups shows that $H^*(C_\alpha; Z)$ has one generator ρ in dimension n and one generator σ in dimension $2n$. Then $\rho^2 = \gamma\sigma$; the integer γ is called the Hopf invariant of the map α. Plainly γ is a homotopy invariant and $\gamma = 0$ if n is odd. Hopf, in introducing this invariant, exhibited maps α, for $n = 2, 4, 8$, with Hopf invariant 1.

Much research has been conducted to determine whether there exist values of n and maps $\alpha: S^{2n-1} \to S^n$ of Hopf invariant 1. It is not difficult to show that the Hopf invariant is a homomorphism $\gamma: \pi_{2n-1}(S^n) \to Z$ and that (for n even) there exist elements α with $\gamma(\alpha) = 2$. Thus the question of the existence of an element of Hopf invariant 1 may be replaced by that of the existence of elements of odd Hopf invariant. This suggests that we look at $H^*(C_\alpha; Z_2)$ instead of $H^*(C_\alpha; Z)$. Then, reducing mod 2, $\rho^2 = Sq^n\rho$, where Sq^i is the Steenrod square; so that this further suggests we look at the "stable" problem of whether a space can exist whose mod 2 cohomology has generators ρ, σ in dimensions k, $n+k$ with $Sq^n\rho = \sigma$ (k large). Adem observed that the Steenrod squares are generated by the squaring operations Sq^{2^i}, and thus showed that there is no map of Hopf invariant 1 if n is not a power of 2. Later, Toda proved that if $n = 16$, there is no α of Hopf invariant 1. We shall prove, as an application of K-theory, the following conclusive result which was first proved by Adams, using ordinary cohomology methods:

(5.1) **Theorem.** If there exists a map $\alpha: S^{2n-1} \to S^n$ of Hopf invariant 1, then $n = 2$, 4 or 8.

To prove this theorem we must define special cohomology operations, that is to say natural transformations, of K-theory, known in the literature as the ψ- (or Adams) operations. Furthermore, we need to consider a ring homomorphism $ch: K^*(X) \to H^*(X; Q)$, the Chern character. We develop these topics next.

We have seen that if ξ is a d-vector bundle over a finite dimensional cell-complex X there exists a map $f: X \to G_d$ and a vector bundle γ over G_d, such that $f^*(\gamma) = \xi$. On the other hand, it is known that $H^*(B_U, Z)$ is a polynomial algebra generated by elements $\bar{c}_i \in H^{2i}(B_U; Z)$, one for each non-negative integer i, with $\bar{c}_0 = 1$. Indeed, $H^*(G_d, Z) = Z[\bar{c}_1, \ldots, \bar{c}_d]$ (Husemoller, D.; Fibre Bundles, McGraw-Hill, Chapter 18). The elements $\bar{c}_i$ are called the universal Chern classes.

We define the Chern classes of ξ to be the elements $c_i(\xi) = f^*(\bar{c}_i) \in H^{2i}(X; Z)$. The total Chern class of ξ is defined to be $c(\xi) = 1 + c_1(\xi) + c_2(\xi) + \ldots + c_d(\xi)$. The following properties hold:

(5.2) if ξ_1 and ξ_2 are equivalent bundles over X, $c(\xi_1) = c(\xi_2)$;

(5. 3) if $f:Y \to X$ is a given map and ξ is a vector bundle over X, $f^*(c(\xi)) = c(f^*(\xi))$;

(5. 4) (Whitney duality formula) if ξ, η are vector bundles over X, then $c(\xi \oplus \eta) = c(\xi)\ c(\eta)$ (Husemoller, D. ; Fibre Bundles, McGraw-Hill, Chapter 16);

(5. 5) **Theorem.** If ξ and η are stably equivalent vector bundles over a space $X \in \mathscr{C}$ then $c(\xi) = c(\eta)$.

Proof. Since the bundles ξ and η are stably equivalent, there exist trivial bundles ϵ_n and ϵ_m so that the vector bundle $\xi \oplus \epsilon_n$ is equivalent to $\eta \oplus \epsilon_m$. By (5. 2) and (5. 4) $c(\xi)\ c(\epsilon_n) = c(\eta)\ c(\epsilon_m)$. But for any trivial bundle ϵ, $c(\epsilon) = 1$, since the associated map $f:X \to B_U$ is nullhomotopic.

Theorem (5. 5) shows that the total Chern classes can be defined on $\tilde{K}(X)$.

The total Chern class of a d-vector bundle ξ over X can be written formally as

$$c(\xi) = (1 + \gamma_1)(1 + \gamma_2) \dots (1 + \gamma_d) ,$$

where each γ_i is a formal 2-dimensional element, $i = 1, \dots, d$, so that $c_i(\xi)$ is equal to the ith elementary symmetric polynomial σ_i in the variables $\gamma_1, \dots, \gamma_d$. We then define the Chern character of ξ by

$$ch(\xi) = e^{\gamma_1} + \dots + e^{\gamma_d} .$$

This makes good sense because ch(ξ) is then effectively a symmetric polynomial in the γ_i, and hence uniquely expressible as a polynomial in the $c_i(\xi)$ with rational coefficients.

Recall that the symmetric polynomials $p_k = \gamma_1^k + \ldots + \gamma_d^k$ satisfy the Newton identities

$$(5.6) \qquad p_k - \sigma_1 p_{k-1} + \sigma_2 p_{k-2} - \ldots + (-1)^k k\sigma_k = 0 ;$$

now we can write p_k just as a function $s_k(\sigma_1, \ldots, \sigma_d)$, called the kth Newton polynomial. Hence, if one writes e^{γ_i} ($i = 1, \ldots, d$) as an infinite power series in γ_i, the Chern character assumes the form

$$(5.7) \qquad ch(\xi) = d + \sum_{k=1}^{\infty} \frac{s_k(c_1(\xi), \ldots, c_d(\xi))}{k!} .$$

This formula shows that ch(ξ) belongs to the graded algebra $\bigoplus_{i \geq 0} H^{2i}(X;Q)$. Moreover, we observe that because of Theorem (5.5) the Chern character can be defined on K(X). In fact, if ξ and ξ' are stably equivalent, the total Chern classes $c(\xi)$ and $c(\xi')$ coincide as elements of a graded algebra; hence, for every i, $c_i(\xi) = c_i(\xi')$. Thus, the summation part of formula (5.7) coincide for ξ and ξ'. Now, for any $([\xi], n) \in K(X)$, where $[\xi]$ is the stable class of the bundle ξ, set $ch([\xi], n) = n - d + ch(\xi)$. Formula (5.7) and the previous observation show that ch is well defined on $([\xi], n)$.

(5.8) **Theorem.** The Chern character is a ring homomorphism of K(X) into H*(X, Q). Furthermore, ch is a natural transformation of functors, and therefore coincides with the character of the theory in the sense of Chapter 3.

Proof. It follows easily from (5. 3) that ch is a natural transformation. Suppose now that ξ and ξ' are respectively, a d-vector bundle and a d'-vector bundle over X. Then $c(\xi) = (1 + \gamma_1) \dots (1 + \gamma_d)$ and $c(\xi') = (1 + \gamma'_1) \dots (1 + \gamma'_{d'})$; it follows that $c(\xi \oplus \xi') = c(\xi)\ c(\xi') = (1 + \gamma_1) \dots (1 + \gamma_d)$ $(1 + \gamma'_1) \dots (1 + \gamma'_{d'})$ and hence, $ch(\xi \oplus \xi') = e^{\gamma_1} + \dots + e^{\gamma_d} + e^{\gamma'_1} + \dots + e^{\gamma'_{d'}} = ch(\xi) + ch(\xi')$. This shows that $ch\{([\xi], n) + ([\xi'], n')\} = ch([\xi \oplus \xi'], n+n') = ch([\xi], n) + ch([\xi'], n')$.

In order to show that ch is a ring-homomorphism, we use the so-called "Splitting Principle"; this asserts that given a complex vector d-bundle $\xi = (E, p, X)$ there exists a space X_ξ and a map $f: X_\xi \to X$ such that:

(i) $f^*: H^*(X, Z) \to H^*(X_\xi, Z)$ is a monomorphism;

(ii) $f^*(\xi) = \eta_1 \oplus \dots \oplus \eta_d$, where η_i is a complex line bundle (i. e., 1-bundle) over X_ξ, $i = 1, \dots, d$. (Husemoller, D. Fibre Bundles, McGraw-Hill, Chapter 16, Section 5). By this principle, we may assume that both ξ and ξ' are a sum of line bundles and, hence, that $\xi \otimes \xi' = \bigoplus_{i,j} \eta_i \otimes \eta'_j$. Therefore, since ch is additive, $ch(\xi \otimes \xi') = \sum_{i,j} ch(\eta_i \otimes \eta'_j)$. On the other hand, for any line bundle η, $c(\eta) = 1 + c_1(\eta)$ and so, $ch(\eta) = e^{c_1(\eta)}$. Also, if η and η' are line-bundles, $c_1(\eta \otimes \eta') = c_1(\eta) + c_1(\eta')$ as one can check by referring to Theorem 3. 4, Chapter 16, of Husemoller's book. These facts readily imply that $ch(\xi \otimes \xi') = ch(\xi)\ ch(\xi')$.

Notice that the splitting principle enables us to regard the γ_i used in the definition of the Chern character as more than just "formal" elements; they may be regarded as $c_1(\eta_i)$ where $f^*(\xi)$

splits as $\eta_1 \oplus \ldots \oplus \eta_d$. Although we do not prove the splitting principle here we remark that the idea of proof is not difficult. It is sufficient to split off one line bundle and, for this, it is sufficient to exhibit a line-subbundle. Now if ξ is a d-vector bundle over X, let $P(\xi)$ be the total space of the associated projective space bundle with fibre map $f: P(\xi) \to X$. Then $f^*(\xi)$ has total space consisting of pairs (v, ℓ) where $v \in \mathbf{C}^d$ and ℓ is a line through the origin in $\mathbf{C}^d$. The subbundle consists of those pairs (v, ℓ) such that $v \in \ell$. One then shows that $H^*(P(\xi); Z)$ is a free $H^*(X; Z)$-module, establishing property (1) of the splitting principle.

Observe that the Chern character is a natural transformation of cohomology theories and defines a homomorphism of the exact couple $EC(X, K^*)$ where $K^*(X) = K^0(X) \oplus K^1(X)$, into the exact couple $EC(X, H^*)$ of ordinary cohomology (also regarded as graded over Z_2 (see (4.12))) with coefficients in Q. It thus induces a homomorphism of the respective spectral sequences. More precisely, $EC(X, K^*)$ leads to a spectral sequence with $E_0^p = K^*(X_p, X_{p-1})$, $E_1^p = H^p(X; Z)$ and $E_\infty^p = F^p(K^*(X))/F^{p+1}(K^*(X))$, where $F^p(K^*(X)) = \ker(K^*(X) \to K^*(X_{p-1}))$. On the other hand, $EC(X, H^*)$ leads to a trivial spectral sequence; see the proof of Theorem 3.23. In general, $\bar{E}_r^p = H^p(X; Q)$, for $r = 1, 2, \ldots$. By inspection of $ch_{(0)} = ch: E_0^p \to \bar{E}_0^p$ at the cochain level, one can see that

$$ch_{(1)}: H^p(X; Z) \to H^p(X; Q)$$

is just the coefficient homomorphism.

We shall now make the following assumption:

(5.9) the space X is a finite cell-complex and $H_*(X)$ is torsion-free, that is to say, is free as a finitely-generated abelian group.

This special assumption implies that $ch_{(1)}$ is a monomorphism. Since $\bar{d}_1 \, ch_{(1)} = ch_{(1)} \, d_1$ and $\bar{d}_1 = 0$, it follows that $d_1 = 0$. Hence, $E_2^p = H^p(X; Z)$ and thus, $ch_{(2)} = ch_{(1)}$. The argument is repeated to show that $d_2 = d_3 = \ldots = 0$ and so, $E_1^p = E_2^p = \ldots = E_\infty^p$ which implies that $ch_{(\infty)}$; $E_\infty^p \to \bar{E}_\infty^p$ is still the coefficient homomorphism and is a monomorphism. Hence $ch: K^*(X) \to H^*(X; Q)$ induces an isomorphism of $F^p(K^*(X))/F^{p+1}(K^*(X))$ onto $H^p(X; Z)$ viewed as a subgroup of $H^p(X; Q)$. This is called the Integrality Theorem. We extract a part of this theorem of particular importance.

For any $z \in K^*(X)$, write $ch(z) = a_0 + a_1 + \ldots + a_j + \ldots$, with $a_j \in H^j(X; Q)$. Thus if $z \in K(X)$, then $a_j = 0$ if j is odd.

(5.10) **Theorem.** For every element z of $K^*(X)$, the first non-zero component of $ch(z)$ of strictly positive degree is an integral class.

Proof. Let us show that $z \in F^{p+1}(K^*(X))$ if, and only if, $a_0 = \ldots = a_p = 0$. It then follows that if $z \in F^p(K^*(X))$ and $chz = a_p + a_{p+1} + \ldots$, and if $\bar{z}$ represents z in E_∞^p, then $ch_{(\infty)}(\bar{z}) = a_p$, so that a_p is an integral class. We now proceed by induction on p.

Consider the diagram $(p = 0,\ z \in F^1(K^*(X)))$

$$\begin{array}{ccc}
F^1(K^*(X)) & & \bigoplus_{j \geq 1} H^j(X; Q) \\
\downarrow i & & \downarrow i' \\
K^*(X) & \xrightarrow{\ ch\ } & H^*(X; Q) \\
\downarrow q & & \downarrow q' \\
K^*(X)/F^1(K^*(X)) & \xrightarrow{\ ch_{(\infty)}\ } & H^0(X; Q)
\end{array} .$$

If $z \in F^1(K^*(X))$, $q'\, ch(i(z)) = ch_{(\infty)} qi(z) = 0$; thus, there is an element $x \in \bigoplus_{j \geq 1} H^j(X; Q)$ such that $i'(x) = ch(i(z))$. Since i and i' are embeddings, $ch(z) \in \bigoplus_{j \geq 1} H^j(X, Q)$ and, hence, $a_0 = 0$.
If $a_0 = 0$, $q'\, ch(z) = 0$ and since $ch_{(\infty)}$ is monic, $q(z) = 0$ and so $z \in F^1 K(X))$.

Assume now that $z \in F^p(K^*(X))$ if, and only if, $a_0 = \ldots = a_{p-1} = 0$. Take the diagram

$$\begin{array}{ccc}
F^{p+1}(K^*(X)) & & \bigoplus_{j \geq p+1} H^j(X; Q) \\
\downarrow i & & \downarrow i' \\
F^p(K^*(X)) & \dashrightarrow & \bigoplus_{j \geq p} H^j(X; Q) \\
\downarrow q & & \downarrow q' \\
F^p(K^*(X))/F^{p+1}(K^*(X)) & \xrightarrow{\ ch_{(\infty)}\ } & H^p(X; Q)
\end{array} .$$

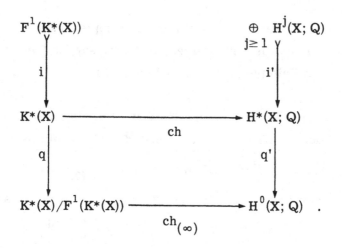

Because of the induction hypothesis, the restriction of ch to $F^p(K^*(X))$ induces a homomorphism into $\bigoplus_{j\geq p} H^j(X, Q)$. A similar argument to that above now shows that ch maps $F^{p+1}(K^*(X))$ into $\bigoplus_{j\geq p+1} H^j(X; Q)$ and that $z \in F^{p+1}(K^*(X))$ if and only if $a_0 = \ldots = a_p = 0$.

The Integrality Theorem has some important consequences which we list below. Recall that the Integrality Theorem and its consequences require assumption (5.9).

(5.11) Let a_p be any integral class of $H^p(X; Q)$. Since $ch_{(\infty)}$ is an isomorphism onto $H^p(X, Z)$, there is a unique $\bar{z} \in F^p(K^*(X))/F^{p+1}(K^*(X))$ such that $ch_{(\infty)}(\bar{z}) = a_p$. Let $z \in F^p(K^*(X))$ be such that $q(z) = \bar{z}$. Because of (5.10), $ch(z) = a_p + \ldots$. Note that z is determined uniquely modulo $F^{p+1}(K^*(X))$.

(5.12) $ch: K^*(X) \to H^*(X; Q)$ is a monomorphism. In fact, if $ch(z) = 0$ and $\dim X = k$, $z \in F^{k+1}(K^*(X)) = 0$.

(5.13) We have seen, at the end of the last chapter, that, X being a finite cell-complex, $K^*(X)$ is of finite type and we computed its rank in terms of the Betti numbers of X; by (5.12) $K^*(X)$ is torsion-free, hence free. Therefore, the Chern character maps $K^*(X)$ isomorphically onto a subring of $H^*(X; Q)$, isomorphic (at least additively) to $H^*(X; Z)$. Of course it does not map $K^*(X)$ onto $H^*(X; Z)$.

We emphasize again that the Integrality Theorem and its consequences hold only when $H_*(X)$ is torsion-free.

Recall from Chapter 4 that the exterior power operations on vector spaces pass to vector bundles over a space X, and hence to B(X). We write these operations $\Lambda^p : B(X) \to B(X)$. As a consequence of a well-known property of vector spaces, they have the following property: for every pair of complex vector bundles over X, say ξ and η, $\Lambda^p(\xi \oplus \eta) = \bigoplus_{q+r=p} \Lambda^q(\xi) \otimes \Lambda^r(\eta)$. Also, if ξ is an n-bundle, $\Lambda^p(\xi) = \epsilon_0$, the trivial zero-bundle, if $p > n$. The Λ-operations will be replaced by ring-operations, which will thus extend to natural transformations of K(X).

For every $z \in B(X)$ and every positive integer k, define

$$\psi_k(z) = s_k(\Lambda^1(z), \ldots, \Lambda^k(z)) ,$$

where s_k is the kth Newton polynomial defined in (5. 6).

It is formal algebra to show that the operations ψ_k are additive. By applying the Splitting Principle in K-theory, it is easy to see that they are also multiplicative, and thus extend to K(X). The ψ-operations are known also as the <u>Adams operations.</u>

(5. 14) **Lemma.** <u>Given</u> $z \in K(X)$ <u>suppose that</u> $ch(z) = a_0 + \ldots + a_j + \ldots$ <u>with</u> $a_j \in H^{2j}(X, Q)$. <u>Then,</u> $ch(\psi_k(z)) = a_0 + ka_1 \ldots + k^j a_j + \ldots$.

Proof. We may write $z = ([\xi], n)$ and compute $ch(\psi_k(\xi))$. Because of the Splitting Principle and since both ch and ψ_k are additive, it is sufficient to take ξ to be a line-bundle η. Then $\psi_k(\eta) = s_k(\Lambda^1(\eta), \ldots \Lambda^k(\eta)) = s_k(\eta, \epsilon_0, \ldots, \epsilon_0) = \eta^k$, and hence, $ch\, \psi_k(\eta) = ch\, \eta^k = (ch\, \eta)^k = e^{c_1(\eta)k} = e^{kc_1(\eta)} = 1 + kc_1(\eta) +$

$k^2(c_1(\eta))^2/2! + \ldots$. Since $\mathrm{ch}\,\eta = e^{c_1(\eta)}$, the proof is complete.

We are now ready to prove the Hopf invariant 1 theorem. Returning to the original formulation, we are led to the following question: does there exist a finite cell-complex X such that $H^*(X, Z) = Z[a]/a^3$, with $\dim a = n$? Observe that because of the remarks that precede (5. 1) we can take n to be an even number 2m. Furthermore, assuming the existence of such a space, $\mathrm{ch}: K^*(X) \to H^*(X, Q)$ is a monomorphism, by (5. 12).

Take the integral class a and let $z \in K(X)$ be such that $\mathrm{ch}(z) = a + \lambda a^2$, with $\lambda \in Q$ (see (5. 11)). Observe that $\mathrm{ch}(z^2) = a^2$. By lemma (5. 14), for every positive integer k, $\mathrm{ch}(\psi_k(z)) = k^m a + k^{2m} \lambda a^2$ and thus,

$$\mathrm{ch}(\psi_k(z) - k^m z) = \lambda k^m (k^m - 1) a^2 .$$

As a consequence of the Integrality Theorem, the number $\tau_k = \lambda k^m (k^m - 1)$ is an integer, for any integer k.

On the other hand, $\mathrm{ch}(\psi_k(z) - k^m z) = \mathrm{ch}(\tau_k z^2)$ implies that $\psi_k(z) - k^m z = \tau_k z^2$ since ch is a monomorphism. In particular, $\psi_2(z) - 2^m z = \tau_2 z^2$ or, according to the definition of ψ_2, $z^2 - 2\Lambda^2(z) - 2^m z = \tau_2 z^2$. From this last equality we get $z^2(\tau_2 - 1) = 2(-\Lambda^2(z) - 2^{m-1} z)$. But z^2 cannot be divisible by 2 in K(X), otherwise a^2 would be divisible by 2 in $H^*(X; Z)$. Therefore $\tau_2 = \lambda 2^m (2^m - 1)$ is an odd integer, and so we can write $\lambda = \dfrac{u}{v \cdot 2^m}$, with u and v odd integers. This shows that the integer $\tau_3 = \lambda\, 3^m(3^m - 1)$ has the form $\tau_3 = \dfrac{u}{v \cdot 2^m} 3^m (3^m - 1)$.

Then, necessarily, 2^m divides $3^m - 1$. We are thus reduced to solving a simple problem in number theory.

Without entering into details it is not difficult to show that if 2^m divides $3^m - 1$, then $m = 1$, 2 or 4, thus establishing Adams' theorem.

We give next some applications of the Hopf invariant 1 theorem of Adams. A space $X \in \mathscr{C}$ has an H-space structure if there is a map $f: X \times X \to X$ such that the diagram

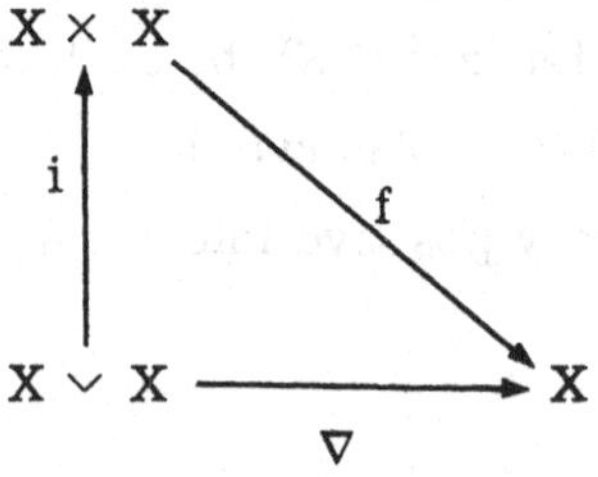

(where ∇ is the folding map and i the inclusion) is homotopy-commutative. The function f is called an H-multiplication.

(5.15) **Theorem.** The only spheres with an H-space structure are S^1, S^3 and S^7.

It is clear that such spheres have an H-space structure, which is induced by the multiplication of complex, quaternion and Cayley numbers. The difficulty is in showing that these are the only spheres with an H-multiplication.

Suppose that $f: S^{n-1} \times S^{n-1} \to S^{n-1}$ is an H-multiplication; we define a map $H(f): S^{2n-1} \to S^n$ as follows. Let S_1 and S_2 be two copies of S^{n-1} and let E_1 and E_2 be n-cells such that $\partial E_i = S_i$, $i = 1, 2$. Then $S^{2n-1} \approx \partial(E_1 \times E_2) = E_1 \times S_2 \cup S_1 \times E_2$;

on the other hand, the suspension ΣS^{n-1} can be viewed as the union of two n-cells E_+ and E_- such that $E_+ \cap E_- = S^{n-1}$. We extend f to $H(f): E_1 \times S_2 \cup S_1 \times E_2 \to \Sigma S^{n-1}$ in such a way that $H(f)(E_1 \times S_2) \subset E_+$ and $H(f)(S_1 \times E_2) \subset E_-$, by looking at E_i as the cone CS_i and setting

$$H(f)(x, t), x') = (f(x, x'), t)_+ \text{ for every } ((x, t), x') \in E_1 \times S_2 \text{ and}$$
$$H(f)(x, (x', t)) = (f(x, x'), t)_- \text{ for every } (x, (x', t)) \in S_1 \times E_2.$$

We now recall that a map $f: S^{n-1} \times S^{n-1} \to S^{n-1}$ has type (d_1, d_2) if its restriction to $S^{n-1} \times p_2$ has degree d_1 and its restriction to $p_1 \times S^{n-1}$ has degree d_2, with $(p_1, p_2) \in S^{n-1} \times S^{n-1}$, base-point. In particular, if f is an H-multiplication, the function f will have type (1, 1).

(5.16) **Lemma.** _If_ $f: S^{n-1} \times S^{n-1} \to S^{n-1}$ _is a map of type_ (d_1, d_2) _the map_ $H(f): S^{2n-1} \to S^n$ _has Hopf invariant_ $d_1 d_2$. (Steenrod, N.; _Cohomology Operations_, Princeton University Press, Chapter 1, Lemma 5.3.)

According to this lemma, if f is an H-multiplication of S^{n-1}, the map H(f) has Hopf invariant 1. Then, by Theorem (5.1), n must be 2, 4 or 8, thus concluding the proof of (5.15).

Now it follows easily that

(5.17) R^n _admits the structure of a real division algebra if, and only if,_ n = 1, 2, 4 _or_ 8. _Indeed_ R^n _admits a non-singular product if and only if_ n = 1, 2, 4 _or_ 8.

A sphere S^n is said to be parallelizable if there is a continuous family of n orthonormal vectors at each point of S^n. Then,

(5.18) S^m is parallelizable if, and only if, $m = 1$, 3 or 7.

In fact, suppose that S^{n-1} is parallelizable; then, to each $x \in S^{n-1}$ we associate a matrix $M(x) = (x, V_1(x), \ldots, V_{n-1}(x))$ of orthonormal vectors; clearly $M(x) \in \mathbf{O}(n)$. Now we define $f: S^{n-1} \times S^{n-1} \to S^{n-1}$ which takes any pair $(x, y) \in S^{n-1} \times S^{n-1}$ into $M(x) \cdot y$. The function f is an H-multiplication in S^{n-1}, thus (5.18) follows from (5.15).

By a vector product on R^n we understand a map $\nu: R^n \times R^n \to R^n$ such that, for every $(x, y) \in R^n \times R^n$,

(i) $\nu(x, y)$ is orthogonal to both x and y;

(ii) $\nu(x, y)^2 = x^2 y^2 - (x \cdot y)^2$.

The usual "vector product" in R^3 is a vector product in this sense; considerations with the Cayley numbers show that R^7 has a vector product. We then prove the following result due to Eckmann.

(5.19) There exists a vector product on R^n if, and only if, $n = 3$ or 7.

If R^n has a vector product ν, embed R^n in R^{n+1} and consider a unit vector b in R^{n+1}, which is orthogonal to R^n. Thus every vector $X \in R^{n+1}$ can be written in a unique way in the form $X = \xi b + x$, with $\xi \in R$ and $x \in R^n$. Now define the product XY in R^{n+1} by associating to a pair $X = \xi b + x$ and

$Y = \eta b + y$ of vectors in R^{n+1}, the vector

$$XY = (\xi\eta - x \cdot y)\, b + \xi y + \eta x + \nu(x, y) .$$

Notice that $X^2 = \xi^2 + x^2$, $Y^2 = \eta^2 + y^2$; moreover, $(XY)^2 = (\xi\eta - x \cdot y)^2 + \xi^2 y^2 + \eta^2 x^2 + \nu(x, y)^2 + 2\xi\eta\, x \cdot y = X^2 Y^2$, by (ii). Therefore, the product XY is norm-preserving, so R^{n+1} admits a non-singular product and $n + 1 = 1, 2, 4$ or 8. (The cases $n = 0, 1$ are trivial.)

It is easy to deduce from (5.19)

(5.20) <u>Among the spheres only S^2 and S^6 admit an almost complex structure.</u>

Appendix: On the Construction of Cohomology Theories *

1. INTRODUCTION

Our object is to describe a procedure for constructing new cohomology theories (on categories of topological spaces) out of given theories. The work summarized below forms part of an investigation in progress which is being carried out jointly with A. Deleanu.[1]

Let $\mathcal{T}$ be the category of based spaces and based maps. A subcategory $\mathcal{T}_1$ of $\mathcal{T}$ is <u>admissible</u> if it is non-empty, full, and closed under the construction of mapping cones, and if, moreover, it contains entire (based) homotopy types. We recall that the suspension is an endofunctor $\Sigma : \mathcal{T} \to \mathcal{T}$ and remark that $\Sigma : \mathcal{T}_1 \to \mathcal{T}_1$ if $\mathcal{T}_1$ is admissible.

A cohomology theory h (see [5]) on the admissible category $\mathcal{T}_1$ is a sequence of functors $h^n : \mathcal{T}_1 \to \mathcal{AB}$, where $\mathcal{AB}$ is the category of abelian groups, and natural transformations $\sigma^n : h^n \to h^{n+1}\Sigma$, $-\infty < n < +\infty$, satisfying the conditions:

(1) if $f_0 \simeq f_1$, then $h^n(f_0) = h^n(f_1)$ (Homotopy axiom)

(2) $\sigma^n : h^n \cong h^{n+1}\Sigma$ (Suspension axiom)

---

* Reproduced from Rend. di Matem. (6) 1, 1968, 219-232.

1 See [I].

(3) If $f:A \to X$ in $\mathcal{T}_1$ and C_f is the mapping cone, with $j:X \to C_f$ the canonical embedding, then

$$h^n(A) \xleftarrow{h^n(f)} h^n(X) \xleftarrow{h^n(j)} h^n(C_f)$$

is exact (Exactness axiom).

If only the homotopy axiom is required then h is a pretheory.

Our programme is to associate with a triple $(\mathcal{T}_1, \mathcal{T}_0, h)$, consisting of two admissible categories with $\mathcal{T}_0 \subseteq \mathcal{T}_1$ and a cohomology theory h on $\mathcal{T}_0$, a pretheory ${}_0h$ on $\mathcal{T}_1$ which agrees with h on $\mathcal{T}_0$ and then to examine conditions on the triple which ensure that ${}_0h$ is a cohomology theory. So far as the note is concerned we will be content to describe the construction of ${}_0h$ and to obtain sufficient conditions on $\mathcal{T}_0$ or the pair $(\mathcal{T}_1, \mathcal{T}_0)$ under which ${}_0h$ is a cohomology theory for all h.

The homotopy class of a map f will be denoted $[f]$.

2. THE CONSTRUCTION OF ${}_0h$

We pass from the categories $\mathcal{T}_0, \mathcal{T}_1$ to the corresponding homotopy categories $\tilde{\mathcal{T}}_0, \tilde{\mathcal{T}}_1$; then $\tilde{\mathcal{T}}_0$ is obtained from $\mathcal{T}_0$ by collecting maps into homotopy classes, so that a morphism of $\tilde{\mathcal{T}}_0$ is a homotopy class of maps. We now carry out the Kan extension procedure[2] for the functor h^n, viewed as a functor from $\tilde{\mathcal{T}}_0$ to $\mathcal{AB}$. Let us suppress the superscript from h^n if no confusion is to be feared. Then, for each X in $|\tilde{\mathcal{T}}_1|$ we

---

[2] This is based on Kan's notion of adjoint functors [3]. We are content to describe it here in a special case, but some very general theorems on Kan extensions have been obtained by F. Ulmer.

consider the <u>category of $\tilde{\mathscr{T}}_0$-objects under</u> X. An object of this category is a morphism $f:X \to Y$ in $\tilde{\mathscr{T}}_1$ with Y in $|\tilde{\mathscr{T}}_0|$ and a morphism $u:f_0 \to f_1$ is a morphism $u:Y_0 \to Y_1$ such that the $\tilde{\mathscr{T}}_1$ diagram

(2.1)

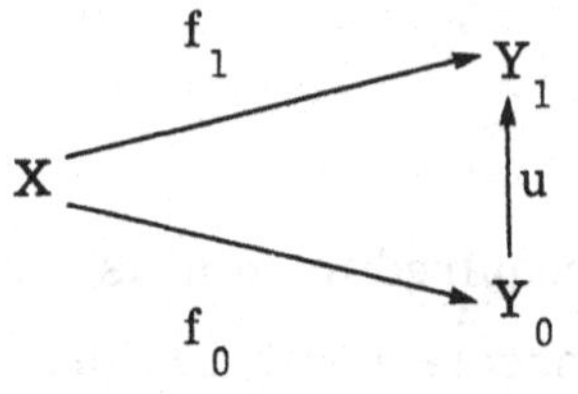

commutes. If we call this category $\tilde{\mathscr{T}}_{10}(X)$, then we define ${}_0h$ as the direct limit of h over $\tilde{\mathscr{T}}_{10}(X)$, or

$$ {}_0h(X) = \varinjlim_{f} (h(Y),\ h(u)) . $$

It is plain from the properties of direct limits that ${}_0h$ is a functor from $\tilde{\mathscr{T}}_1$ to $\mathscr{A}\mathscr{B}$, and that ${}_0h$ coincides with h on $\tilde{\mathscr{T}}_0$. It thus remains to define the natural transformations

$$ {}_0\sigma^n : {}_0h^n \to {}_0h^{n+1}\Sigma . $$

In view of the definition of the direct limit it suffices to define a homomorphism $\sigma_f : h^n(Y) \to {}_0h^{N+1}(\Sigma X)$ corresponding to $f:X \to Y$ in $\tilde{\mathscr{T}}_1$ such that, given (2.1),

(2.2) $$ \sigma_{f_1} = \sigma_{f_0} h^n(u) . $$

We merely set

$$\sigma_f = \iota_{\Sigma f} \circ \sigma^n(Y) , \tag{2.3}$$

where ι_g is the canonical map from $h^{n+1}(Z)$ to the direct limit ${}_0h^{n+1}(\Sigma X)$ corresponding to the index $g:\Sigma X \to Z$. In fact this description of ${}_0\sigma^n$ merely renders explicit in our case the Kan extension procedure, applied to a natural transformation of functors. Plainly ${}_0\sigma^n$ extends σ^n in the obvious sense, and ${}_0h = ({}_0h^n, {}_0\sigma^n, -\infty < n < +\infty)$ is a pretheory on $\mathscr{T}_1$.

Note that the Kan extension procedure described here extends a pretheory on $\mathscr{T}_0$ to a pretheory on $\mathscr{T}_1$; we have made no use of the fact that h is a cohomology theory.

3. THE CONSTRUCTION OF COHOMOLOGY THEORIES

We are going to study conditions on $\mathscr{T}_1$ and $\mathscr{T}_0$ under which we may infer that ${}_0h$ is a theory, given that h is a theory. We suppose that $\mathscr{T}_0$ has the following property I. Given a diagram

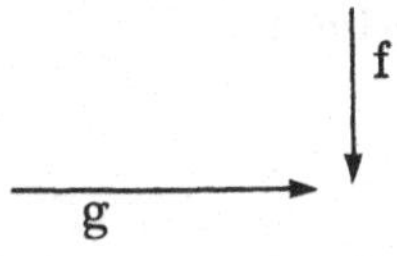

in $\mathscr{T}_0$ where f, g are fibrations then the induced fibration diagram (= pull-back) is also in $\mathscr{T}_0$.

Proposition 3.1. *If $\mathscr{T}_0$ has property I then $\mathscr{T}_0$ is closed with respect to topological products (of pairs of spaces) and to construction of loop spaces.*

Proof. We obtain $X \times Y$ by pulling back

$$\begin{array}{ccc} & & X \\ & & \big\downarrow \\ Y & \longrightarrow & 0 \end{array}$$

Since $\mathscr{T}_0^-$ contains points, it contains the contractible path space EX. Then by pulling back

$$\begin{array}{ccc} & & EX \\ & & \big\downarrow \\ EX & \longrightarrow & X \end{array}$$

we get a space of the homotopy type of the loop space ΩX.

Proposition 3.2. *If $\mathscr{T}_0^-$ has property I then $\tilde{\mathscr{T}}_0^-$ has weak pull-backs, relative to $\tilde{\mathscr{T}}^-$.*

Proof. The assertion is that given

$$\begin{array}{ccc} & & \\ & & \big\downarrow \phi \\ & \xrightarrow{\ \psi\ } & \end{array}$$

in $\tilde{\mathscr{T}}_0$, there exists a commutative square

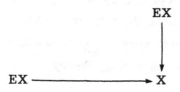

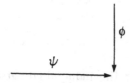

$$\begin{array}{ccc} \cdot & \xrightarrow{\alpha} & \cdot \\ \beta \big\downarrow & & \big\downarrow \phi \\ \cdot & \xrightarrow{\psi} & \cdot \end{array}$$

in $\tilde{\mathscr{T}}_0^-$ such that, for any γ, δ in $\tilde{\mathscr{T}}^-$ with $\phi\gamma = \psi\delta$, there exists ρ in $\tilde{\mathscr{T}}^-$ with $\alpha\rho = \gamma$, $\beta\rho = \delta$. This was essentially proved in [2]. The key idea is to choose $f: A \to X$, $g: B \to X$ in the classes ϕ, ψ, to factorize f in $\mathscr{T}_0^-$ as $A \xrightarrow{r} E_f \xrightarrow{f'} X$, where r is a homotopy equivalence and f' is a fibration, to factorize g similarly as $B \xrightarrow{s} E_g \xrightarrow{g'} X$ and then apply property I to

$$\begin{array}{ccc} & & E_f \\ & & \big\downarrow f' \\ E_g & \xrightarrow{g'} & X \end{array}$$

to get

$$\text{(3.3)} \qquad \begin{array}{ccc} T & \xrightarrow{a'} & E_f \\ b' \big\downarrow & & \big\downarrow f' \\ E_g & \xrightarrow{g'} & X \end{array}$$

Then $\alpha = [\bar{r}a']$, $\beta = [\bar{s}b']$, where $\bar{r}$, $\bar{s}$ are homotopy inverses of r, s. Plainly

$$\phi\alpha = [f\bar{r}a'] = [f'r\bar{r}a'] = [f'a'], \quad \psi\beta = [g'b'], \text{ similarly,}$$

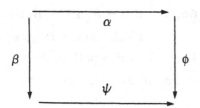

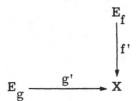

so $\phi\alpha = \psi\beta$. Moreover if $\gamma = \psi\delta$, let $k:U \to A$, $l:U \to B$ be in the classes γ, δ, so that $fk \simeq gl$ or $f'rh \simeq gl$. Since f' is a fibration $rk \simeq t$ with $f'\,t = gl = g'\,sl$. Since (3.3) is a pull-back, there exists $m:U \to T$ with $a'\,m = t \simeq rk$, $b'\,m = sl$. Then, if $\rho = [m]$,

$$\alpha\rho = [\bar{r}a'm] = [\bar{r}rk] = [k] = \gamma, \quad \beta\rho = \delta, \text{ similarly.}$$

We revert now to our cohomology h on $\mathcal{T}_0^-$. We fix a space X in $\mathcal{T}_1$.

Theorem 3.4. Consider the set of pairs (α, f) where[3] $f:X \to Y$ in $\tilde{\mathcal{T}}_1^-$ with Y in $\mathcal{T}_0^-$ and $\alpha \in h(Y)$. Introduce the relation $(\alpha_1, f_1) \sim (\alpha_2, f_2)$ if there is a diagram

(3.5) 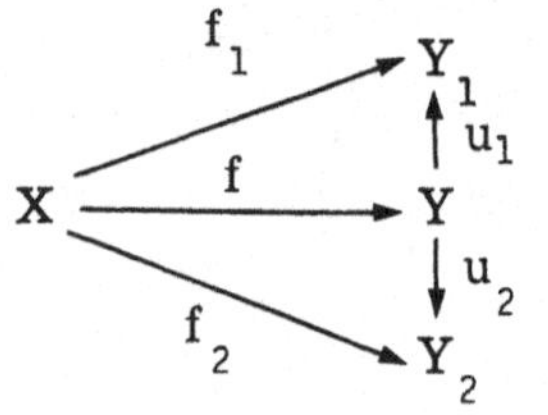

with u_1, u_2 in $\tilde{\mathcal{T}}_0$, $u_1 f = f_1$, $u_2 f = f_2$, and $h(u_1)\alpha_1 = h(u_2)\alpha_2$.
If $\mathcal{T}_0$ has property I then this relation is an equivalence relation.

[3] Note that f now stands for a homotopy class of maps, so that the notation $[f]$ is unnecessary.

Proof. Only transitivity is in question. Suppose given (3.5) and

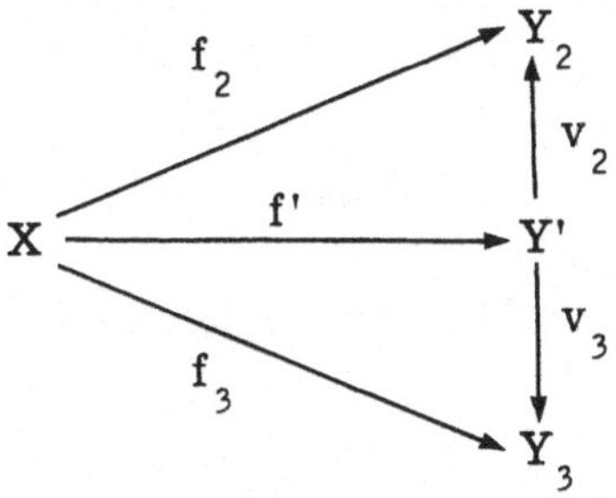

in $\tilde{\mathcal{T}}_1$, with v_2, v_3 in $\tilde{\mathcal{T}}_0$, $v_2 f' = f_2$, $v_3 f' = f_3$, $h(v_2)\alpha_2 = h(v_3)\alpha_3$, so that $(\alpha_2, f_2) \sim (\alpha_3, f_3)$. We have the commutative diagram

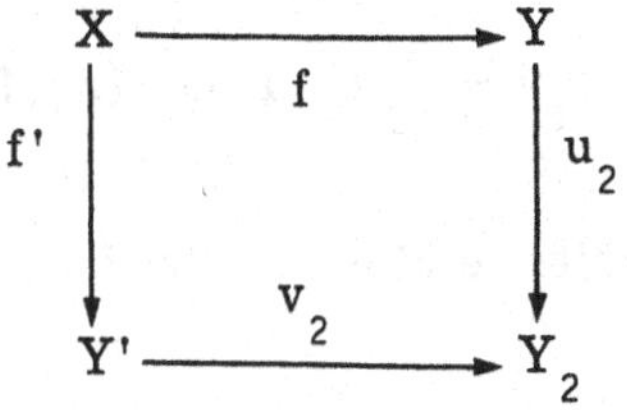

and use Proposition 3.2 to construct the commutative diagram

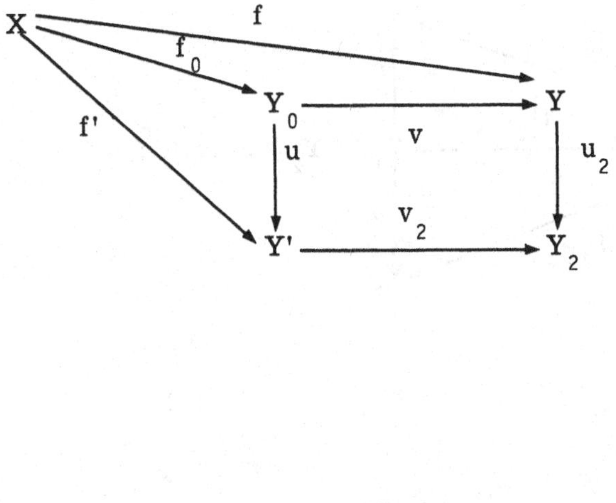

with u, v in $\tilde{\mathcal{T}}_0$. From this we construct

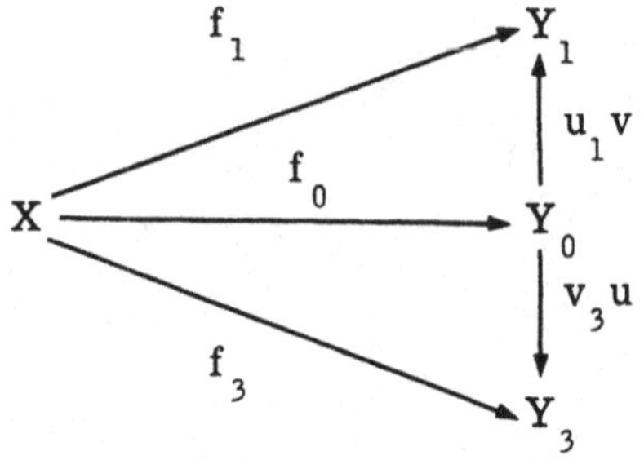

and observe that $u_1 v f_0 = u_1 f = f_1$, $v_3 u f_0 = v_3 f' = f_3$, and

$$h(u_1 v)\alpha_1 = h(v)h(u_1)\alpha_1 = h(v)h(u_2)\alpha_2 ,$$

$$h(v_3 u)\alpha_3 = h(u)h(v_3)\alpha_3 = h(u)h(v_2)\alpha_2 ,$$

so that $h(u_1 v)\alpha_1 = h(v_3 u)\alpha_3$, since $u_2 v = v_2 u$. Thus $(\alpha_1, f_1) \sim (\alpha_3, f_3)$.

We write $[\alpha, f]$ for the equivalence class of (α, f).

Proposition 3. 6. *Define* $(\alpha_1, f_1) + (\alpha_2, f_2)$ *to be* $(h(p_1)\alpha_1 + h(p_2)\alpha_2, \{f_1, f_2\})$.

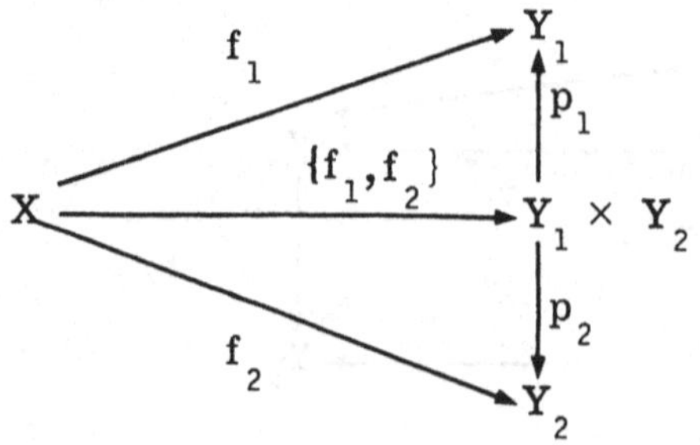

This addition induces an abelian group structure in the set E of equivalence classes $[\alpha, f]$; the resulting group is $\varinjlim_{f} (h(Y), h(u))$.

Proof. It is straightforward to verify that the addition is compatible with the equivalence relation (notice that the addition is defined because $\tilde{\mathcal{T}}_0$ is closed under topological products), and that the set of equivalence classes thereby forms an abelian group. We particularly notice that $[\alpha, f] = 0$ if and only if there is a diagram

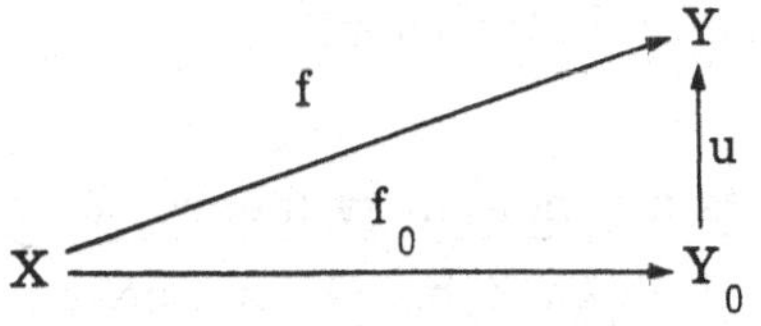

with $uf_0 = f$, u in $\tilde{\mathcal{T}}_0$, and $h(u)\alpha = 0$; also $-[\alpha, f] = [-\alpha, f]$, $[\alpha + \beta, f] = [\alpha, f] + [\beta, f]$.

Define $\iota_f : h(Y) \to E$ by $\iota_f(\alpha) = [\alpha, f]$. Then ι_f is a homomorphism and, given (2.1), $\iota_{f_1} = \iota_{f_0} h(u)$. It is now easy to see that E, together with the homomorphisms ι_f, enjoys the universal mapping property which characterizes the direct limit $\varinjlim_{f} (h(Y), h(u))$.

We have thus proved that

$$(3.7) \qquad {}_0h(X) = E = \varinjlim_{f} (h(Y), h(u)) .$$

With the interpretation of ${}_0h(X)$ given by Theorem 3.6 it now remains to identify the suspension mapping

$$ {}_0\sigma^n : {}_0h^n \to {}_0h^{n+1}\Sigma \,. $$

In view of (2.3) and the definition of ι_f just given, it is clear that ${}_0\sigma$ is defined by

$$ (3.8) \qquad {}_0\sigma[\alpha, f] = [\sigma\alpha, \Sigma f]\,. $$

We have thus expressed the pretheory $({}_0h, {}_0\sigma)$ in a particularly simple and explicit form in terms of the pretheory (h, σ). We now state and prove the main theorem.

Theorem 3.9. <u>If</u> (h, σ) <u>is a cohomology theory on</u> $\mathcal{T}_0^-$ <u>and if</u> $\mathcal{T}_0^-$ <u>has property</u> I, <u>then</u> $({}_0h, {}_0\sigma)$ <u>is a cohomology theory on</u> $\mathcal{T}_1$, <u>for any admissible</u> $\mathcal{T}_1 \supseteq \mathcal{T}_0^-$.

Proof. We must verify the suspension and exactness axioms for $({}_0h, {}_0\sigma)$, given that they are satisfied by (h, σ). We deal first with the suspension axiom and prove that the homomorphism ${}_0\sigma$ of (3.8) is an isomorphism.

For any Y in $|\mathcal{T}_0^-|$, let

$$ (3.10) \qquad d: Y \to \Omega\Sigma Y, \quad e: \Sigma\Omega Y \to Y $$

be the natural transformations. Then d, e are in $\mathcal{T}_0^-$ by Proposition 3.1. Let us indicate by f' the adjoint map paired to f under the adjunction

$$\mathscr{T}(\Sigma X,\ Y) \cong \mathscr{T}(X,\ \Omega Y),\ X,\ Y \in \mathscr{T};$$

here f belongs to $\mathscr{T}(\Sigma X,\ Y)$, f' to $\mathscr{T}(X,\ \Omega Y)$. Then $d = 1'$ and

$$(3.11) \qquad f = e \circ \Sigma f', \qquad f{:}\Sigma X \to Y,$$

$$(3.12) \qquad (\Sigma f)' = d \circ f, \qquad f{:}X \to Y.$$

Now consider $[\alpha, f] \in {}_0h^{n+1}(\Sigma X)$, $f{:}\Sigma X \to Y$. Let $\beta \in h^n(\Omega Y)$ be such that $\sigma\beta = h(e)\alpha$. Then, by (3.11),

$$[\alpha, f] = [h(e)\alpha, \Sigma f'] = [\sigma\beta, \Sigma f'] = {}_0\sigma[\beta, f'] .$$

This proves that ${}_0\sigma$ is onto. To prove ${}_0\sigma$ one-one, let $[\alpha, f] \in {}_0h^n(X)$, $f{:}X \to Y$, and suppose $[\sigma\alpha, \Sigma f] = 0$. This means that there exists a diagram (commutative in $\tilde{\mathscr{T}}_1$)

$$(3.13) \qquad \begin{array}{ccc} \Sigma X & \xrightarrow{\ \Sigma f\ } & \Sigma Y \\ & {\searrow}^{g_0} & \uparrow u_0 \\ & & Y_0 \end{array}$$

with Y_0 in $\mathscr{T}_0$, $h(u_0)\sigma\alpha = 0$. Diagram (3.13) gives rise to the commutative diagram

$$(3.14) \qquad \begin{array}{ccc} \Sigma X & \xrightarrow{\ \Sigma f\ } & \Sigma Y \\ & {\searrow}^{\Sigma f_1} & \uparrow u \\ & & \Sigma Y_1 \end{array}$$

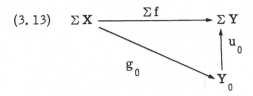

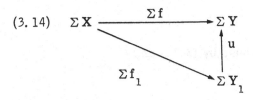

where $Y_1 = \Omega Y_0$ (so that ΣY_1 is in $\mathscr{T}_0^-$), $u = u_0 e$, $f_1 = g_0'$; the diagram commutes by (3.11). Moreover, $h(u)\sigma\alpha = 0$.

We have the further diagrams

(3.15)

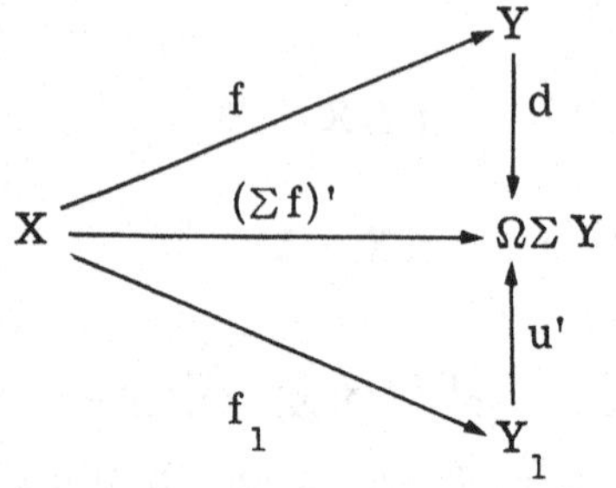

$$(3.16) \quad \Sigma Y_1 \xrightarrow{\Sigma u'} \Sigma \Omega \Sigma Y \underset{e}{\overset{\Sigma d}{\underset{\longrightarrow}{\longleftarrow}}} \Sigma Y .$$

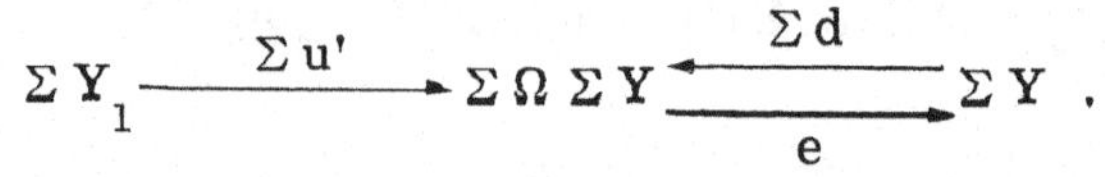

Then (3.15) is commutative; the commutativity of the upper triangle is just (3.12); that of the lower triangle follows by applying adjoints to (3.14). In (3.16) we have

$$(3.17) \quad e\Sigma u' = u , \qquad (3.11)$$

$$(3.18) \quad e\Sigma d = 1 \qquad (3.11, \text{ since } d = 1') .$$

Let $\beta \in h^n(\Omega\Sigma Y)$ be such that $\sigma\beta = h(e)\sigma\alpha$. Then $h(\Sigma d)\sigma\beta = \sigma\alpha$ by (3.18), so that

$$(3.19) \quad h(d)\beta = \alpha ,$$

since σ is one-one. Thus, by (3.15) and (3.19)

$$[\alpha, f] = [\beta, (\Sigma f)'] = [h(u')\beta, f_1]\,.$$

Now, by (3.17) and by hypothesis, $h(\Sigma u')(\sigma\beta) = h(\Sigma u')h(e)\sigma\alpha = h(u)\sigma\alpha = 0$, so that, σ being one-one, $h(u')\beta = 0$. Thus $[\alpha, f] = [0, f_1] = 0$, and the proof that ${}_0\sigma$ is one-one is complete.

We now prove that ${}_0h$ satisfies the exactness axiom. We suppose given

$$A \xrightarrow{\;g\;} X \xrightarrow{\;j\;} C_g$$

in $\mathcal{T}_1^-$; here we should insist that g is a map, that is, in $\mathcal{T}_1^-$ rather than $\tilde{\mathcal{T}}_1^-$. Let $[\alpha, f] \in {}_0h(X)$, $f: X \to Y$, and suppose that ${}_0h(g)[\alpha, f] = 0$. This means that we have a diagram

$$(3.20) \qquad \begin{array}{ccccc} A & \xrightarrow{\;g\;} & X & \xrightarrow{\;j\;} & C_g \\ \downarrow{\scriptstyle f_0} & & \downarrow{\scriptstyle f} & & \\ Y_0 & \xrightarrow{\;u\;} & Y & \xrightarrow{\;k\;} & C_u \end{array}$$

with u in $\mathcal{T}_0^-$, $fg \simeq uf_0$, and $h(u)\alpha = 0$; moreover if k is the natural embedding of Y in the mapping cone of u, then k is in $\mathcal{T}_0^-$. We infer from (3.20) a map $f_1: C_g \to C_u$ with $f_1 j \simeq kf$. Since h satisfies the exactness axiom, there exists $\beta \in h(C_u)$ with $h(k)(\beta) = \alpha$. Then

$${}_0h(j)[\beta, f_1] = [\beta, f_1 j] = [\beta, kf] = [\alpha, f]\,.$$

Thus ${}_0h$ satisfies the exactness axiom and Theorem 3.9 is completely proved. Notice that effectively Theorem 3.9 gives conditions under which h extends to a cohomology theory ${}_0h$ on the whole of $\mathcal{T}$.

Examples 3.21. (i) Let $\mathcal{T}_1 = \mathcal{T}$, $\mathcal{T}_0 = \mathcal{C}$, category of spaces of the homotopy type of CW-complexes. We may then describe the process of passing from h to ${}_0h$ as the Čech extension of the theory h.

(ii) $\mathcal{T}_1 = \mathcal{T}$, $\mathcal{T}_0 = \mathcal{C}(\chi_0)$, category of spaces of the homotopy type of countable CW-complexes.

In order further to extend the applicability of the extension process we should generalize property I. Examination of the proof of Theorem 3.9 shows that we need $\mathcal{T}_0$ in fact to have the following properties:

(a) $\mathcal{T}_0$ is closed under products (of two objects).

(b) Let $\tilde{\Omega}$ be right adjoint to Σ tn $\mathcal{T}_1$ (see [3]); then $\mathcal{T}_0$ is closed under $\tilde{\Omega}$.

(c) $\mathcal{T}_0$ has weak pull-backs relative to $\mathcal{T}_1$.

Let us say that $\mathcal{T}_0$ has property II (relative to $\mathcal{T}_1$) if it enjoys properties (a), (b), (c) above. Then, as a corollary of the proof of Theorem 3.9 we have

Theorem 3.22. If (h, σ) is a cohomology theory on $\mathcal{T}_0$ and if $\mathcal{T}_0$ has property II relative to $\mathcal{T}_1$ then $({}_0h, \sigma_0)$ is a cohomology theory on $\mathcal{T}_1$.

Notice that Theorem 3.22 may be regarded as a relativization of Theorem 3.9, whereby we impose conditions on the pair $(\mathcal{T}_1, \mathcal{T}_0)$ rather than just on $\mathcal{T}_0$; for $\mathcal{T}_0$ certainly has property II relative to $\mathcal{T}$ if it has property I.

To give one set of applications of Theorem 3.22 we take to be $\mathcal{T}_s$, the category of 1-connected spaces.[4] Then $\tilde{\Omega} X$, X in $\mathcal{T}_s$, is just the universal cover of ΩX, the loop space on X. Given a diagram

(3.23)
$$\begin{array}{ccc} & & Y_0 \\ & & \downarrow{\scriptstyle f} \\ Y_1 & \xrightarrow{g} & Y \end{array}$$

in $\mathcal{T}_0$, where f, g are fibrations, construct the induced fibration diagram

(3.24)
$$\begin{array}{ccc} Z_0 & \xrightarrow{g'_0} & Y_0 \\ {\scriptstyle f'_0}\downarrow & & \downarrow{\scriptstyle f} \\ Y_1 & \xrightarrow{g} & Y \end{array}$$

Then Z_0 is 0-connected. Let Z be the universal cover of Z_0, with cover map $p: Z \to Z_0$ and let

(3.25)
$$\begin{array}{ccc} Z & \xrightarrow{g_0} & Y_0 \\ {\scriptstyle f_0}\downarrow & & \downarrow{\scriptstyle f} \\ Y_1 & \xrightarrow{g} & Y \end{array}$$

---

[4] We should, strictly speaking, restrict the spaces in $\mathcal{T}_s$ so that they do have universal covers. This can be achieved by only considering spaces which are locally path-connected and locally simply-connected in the weak sense.

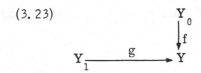

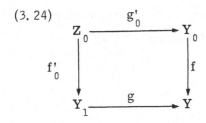

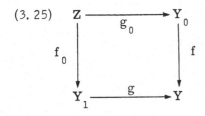

be the diagram with $f_0 = f'_0 p$, $g_0 = g'_0 p$. We call (3.25) the <u>1-connected pull-back</u> of (3.23). We say that $\mathcal{T}_0$ <u>has property</u> I_1 if (3.25) is in $\mathcal{T}_0$, that is, if $\mathcal{T}_0$ is closed under 1-connected pull-backs.

Proposition 3.26. <u>If</u> $\mathcal{T}_0$ <u>has property</u> I_1, <u>then it has property</u> II <u>relative to</u> $\mathcal{T}_S$.

Proof. Let $\mathcal{T}_0$ have property I_1. Notice that (3.25) coincides with the induced fibration diagram if the pull-back space Z_0 is 1-connected. Thus it is evident that $\mathcal{T}_0$ has property (a). Also the argument of Proposition 3.1 may be reproduced here to show that $\mathcal{T}_0$ is closed under $\tilde{\Omega}$. Finally we prove property (c). The argument of Proposition 3.2 establishes that, given

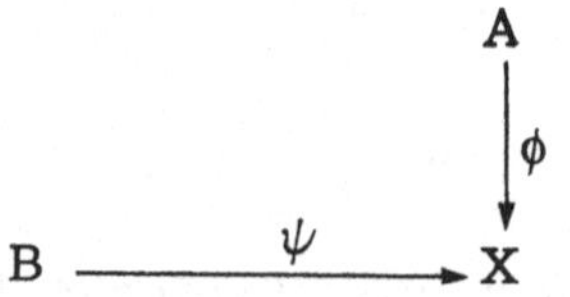

in $\mathcal{T}_0$, there is a square

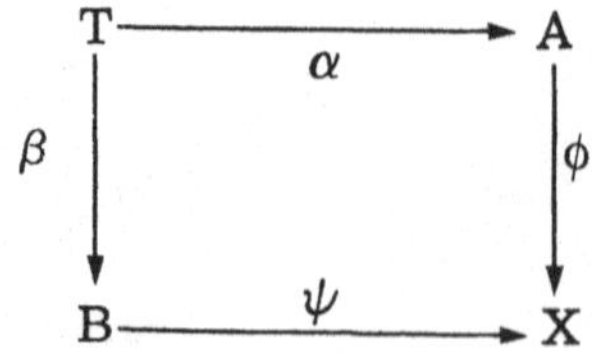

in $\mathcal{T}$ satisfying the weak pull-back condition for any $\gamma : U \to A$, $\delta : U \to B$ in $\mathcal{T}$ with $\phi\gamma = \psi\delta$. Moreover T is the pull-back space of a pair of fibrations in $\mathcal{T}_0$. Thus if T_1 is the universal cover of T and $p:T_1 \to T$ the covering map then

$$\begin{array}{ccc} T_1 & \xrightarrow{\alpha_1} & A \\ {\scriptstyle \beta_1}\downarrow & & \downarrow{\scriptstyle \phi} \\ B & \xrightarrow{\psi} & X \end{array}$$

where $\alpha_1 = \alpha[p]$, $\beta_1 = \beta[p]$, is a commutative square in $\mathcal{T}_0$, and it plainly satisfies the weak pull-back condition for any $\gamma:U \to A$, $\delta:U \to B$ with $\phi\gamma = \psi\delta$ and U 1-connected. This proves the proposition.

Corollary 3.27. *Let $\mathcal{T}_0$ be an admissible subcategory of the category $\mathcal{T}_s$ of 1-connected spaces. If $\mathcal{T}_0$ has property I_1, then, for any cohomology theory h on $\mathcal{T}_0$ the pretheory ${}_0h$ on $\mathcal{T}_s$ is also a cohomology theory.*

4. APPLICATIONS

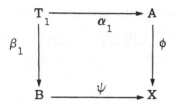

We apply Corollary 3.27 to the case where we take $\mathcal{T}_0$ to be the subcategory of $\mathcal{T}_s$ consisting of spaces whose homotopy groups belong to a class C in the sense of Serre [4]. Since all spaces in $\mathcal{T}_s$ are 1-connected it follows that the homology groups of spaces in $\mathcal{T}_0$ (in positive dimensions) also belong to the class C; indeed, the two conditions are equivalent. It is plain that if $f:X \to Y$ with X, Y in $\mathcal{T}_0$ then the mapping cone C_f is also in $\mathcal{T}_0$; for it is certainly 1-connected and its homology groups inherit from X and Y the property of belonging to C - it is only necessary to look at the exact homology sequence. Thus $\mathcal{T}_0$, which we write $\mathcal{T}_0(C)$, is an admissible category.

Theorem 4.1. *Let* C *be a class of abelian groups in the sense of Serre and let* $\mathcal{T}_0(C)$ *be the subcategory of* $\mathcal{T}_S$ *consisting of spaces whose homotopy groups belong to* C. *Then any cohomology theory* h *on* $\mathcal{T}_0(C)$ *extends to a theory* ${}_0h$ *on* $\mathcal{T}_S$.

Proof. In the light of Corollary 3.27 it is sufficient to verify that $\mathcal{T}_0(C)$ has property I_1. We complete the diagram (3.24) to

$$\begin{array}{ccc} F & = & F \\ \downarrow & & \downarrow \\ Z_0 & \longrightarrow & Y_0 \\ {\scriptstyle f'_0}\downarrow & & \downarrow{\scriptstyle f} \\ Y_1 & \longrightarrow & Y \end{array}$$

where F is the fibre of f and of f'_0. It follows from the exact homotopy sequence for f that $\pi_i(F) \in C$, $i \geq 1$. It then follows from the exact homotopy sequence for f'_0 that $\pi_i(Z_0) \in C$, $i \geq 1$. Since Z is the universal cover of Z_0 it follows that $\pi_i(Z) \in C$, $i \geq 2$ and Z is 1-connected. Thus $Z \in \mathcal{T}_0(C)$ so that property I_1 is verified.

We recall that the following are examples of Serre classes of abelian groups.

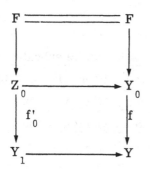

(i) The class of finite abelian groups.

(ii) The class of finitely-generated abelian groups.

(iii) Let P be a family of prime numbers. Then $A \in C$ if and only if A contains no elements whose orders are in P.

We plan later to consider the nature of the cohomology theories which arise by applying Theorem 4. 1. We close this paper with the remark that any cohomology theory h on $\mathcal{T}_S$ can be extended in a trivial way to $\mathcal{T}$; namely, we set, for any space X,

$$h^n(X) = h^{n+2}(\Sigma^2 X) \quad .$$

This device would, however, destroy* any multiplicative structure in h; we have not discussed multiplicative structure (and operations) in this paper, but they plainly play a vital role in any cohomology theory. We will show (elsewhere) that the Kan extension process preserves multiplicative structure.

---

* (Added later) Actually, this device does not destroy multiplicative structure! The point which I overlooked is that, although we suspend X, we apply the diagonal map in X, not in $\Sigma^2 X$.

BIBLIOGRAPHY

[1] A. Deleanu and P. J. Hilton: 'Some remarks on general cohomology theories', Math. Scand. 22 (1968), 227-240.

[2] B. Eckmann and P. J. Hilton: 'Unions and intersections in homotopy theory', Comm. Math. Helv. 38 (1964), 293-307.

[3] D. M. Kan: 'Adjoint functors', Trans. Amer. Math. Soc. 87 (1958), 294-329.

[4] E. H. Spanier: Algebraic Topology, McGraw Hill (1966).

[5] G. Whitehead: 'Generalized homology theories', Trans. Amer. Math. Soc. 102 (1962), 227-283.